Alfons Vogelpohl
Distillation

———

The Theory

2nd, revised Edition

DE GRUYTER

Author
Professor
Dr.-Ing. Alfons Vogelpohl
Clausthal University of Technology
Institute for Separation and Process Technology
Leibnizstr. 15
38678 Clausthal-Zellerfeld
Germany

ISBN 978-3-11-073972-5
e-ISBN (PDF) 978-3-11-073973-2
e-ISBN (EPUB) 978-3-11-073981-7

Library of Congress Control Number: 2021931835

Bibliographic information published by the Deutsche Nationalbibliothek
The Deutsche Nationalbibliothek lists this publication in the Deutsche Nationalbibliografie;
detailed bibliographic data are available on the internet at http://dnb.dnb.de.

MIX
Papier aus verantwor-
tungsvollen Quellen
FSC
www.fsc.org   FSC® C083411

Nothing is more practical than a good theory.
Kurt Lewin

Preface

Distillation is the most important technical process for the separation of liquid or gaseous mixtures into fractions or the components like the recovery of petroleum compounds from crude oil or the separation of air into oxygen and nitrogen. It is surprising, therefore, that despite the numerous papers dealing with distillation published since the end of the 19th century there is no comprehensive theory of the distillation process.

This book, based on more than 40 years of teaching courses on Mass Transfer Processes intends to fill this gap starting from the basic equation of ternary distillation published by Hausen in 1935 and by fully exploiting the remarkable properties of this equation covering all modes of distillation.

The first three chapters describe the modes of distillation and develop the basic equations governing the distillation process. Chapter 4 is the central chapter of the book discussing in detail the binary up to multicomponent distillation of ideal mixtures. Chapter 5 proves that the results of Chapter 4 based on ideal mixtures are valid qualitatively also for real mixtures including mixtures with azeotropes.

Special attention has been given to interpret the coefficients and parameters of the theory in relation to the physical phenomena of the distillation process. Numerous figures and examples serve to illustrate the analytical relationships.

The book is intended as a graduate or postgraduate textbook for an advanced course on distillation but will also help the practicing engineer to better understand the complex interrelationships of multicomponent distillation.

I dedicate the book to my Ph. D. supervisor Professor H. Hausen, who taught me the basics of academic research, and Professor E.-U. Schlünder, who initiated my return from the industry to the university. Without them this book would never have been written.

The author gratefully acknowledges the efficient support of Mrs. Karin Sora, Mrs. Julia Lauterbach and Mr. Hannes Kaden at De Gruyter in editing this book.

January, 2015

Alfons Vogelpohl
Clausthal-Zellerfeld, Germany

https://doi.org/10.1515/9783110739732-201

Contents

Introduction

Distillation developed in ancient times as an art and was used throughout the middle ages e.g. in the production of perfumes like rose oil or medicines from herbs [1]. In the 19th century it was developed into an industrial operation to take care of the growing needs for higher purity raw materials and products. This required a better understanding of the distillation process resulting in more efficient apparatus and new applications like the separation of air into oxygen and nitrogen, the production of a variety of products from petroleum or high purity intermediates required for the production of polymers or until then unknown chemical compounds [2].

The scientific understanding of the distillation process started in 1893 with Hausbrand [3] and Sorel [4] who proposed the "Theoretical Stage Concept" which even today is widely applied in the design of distillation columns and the interpretation of the distillation process. In 1922 the "Mass Transfer Concept" was introduced by Lewis [5]. The latter concept is the first systematic application of Chemical Engineering Principles to the design and operation of distillation plants with the emphasis on the separation of binary mixtures.

Even though the "Theoretical Stage Concept" and the "Mass Transfer Concept" can be applied to the design of multicomponent distillation columns, the exponential growth of the numerical calculations with an increasing number of components restricted the application of the concepts practically to the separation of three component mixtures. It was for ideal systems only that analytical solutions were developed like the Rayleigh equation [6] dealing with the simple distillation, the Fenske equation [7, 8] for the number of theoretical stages at total reflux, the equations of Underwood [9, 10] giving the minimum reflux and the feasible products for any number of components and the fundamental papers of Hausen [11, 12] dealing with three component mixtures and the role of the separation lines in distillation. The author extended the findings of Fenske, Underwood and Hausen into an improved theory of the distillation of ideal mixtures [13, 14], which can also be applied to the distillation of real mixtures of any complexity at least approximately [15, 16].

The theory of distillation as presented in this book is based on the "Mass Transfer Concept" as this is the fundamental concept encompassing all modes of distillation and as it allows to visualize the physical background of distillation in the simplest way possible and to comprehend the interrelations between the distillation of ideal and real mixtures.

Numerous graphs and problems serve to illustrate the different modes of distillation and to reveal their interrelations.

Even though efficient numerical computer methods are available nowadays for calculating the distillation of complex real multicomponent mixtures, the theory described in this book in contrast provides for a comprehensive understanding of the complexities of multicomponent distillation and thus is useful not only for the teacher

https://doi.org/10.1515/9783110739732-202

in his lectures on distillation but also for the practicing chemical engineer faced with multicomponent distillation problems.

All equations required for solving a specific distillation problem are given in the related Chapters of the book with the derivation of the more complex equations discussed in the Appendices. In addition all essential design elements are implemented in MATLAB programs presenting the results in numerical and/or graphical form.

1 The principles and modes of distillation

The principle of distillation is based on the thermodynamic property of most liquid mixtures that the vapour produced from a boiling mixture has a composition enriched in the lower boiling components of the liquid which allows to separate a liquid mixture into fractions with compositions different from the liquid mixture and even into its components. The most basic devices of implementing a distillation are the discontinuous or batchwise distillation via the simple distillation or the continuous flash distillation.

1.1 Simple distillation

Figure 1.1 shows a set-up to carry out such a "simple distillation". It consists of a heated still-pot, a condenser to liquefy the vapour produced in the still-pot and a receiver to collect the distillate. If heat is continuously added to the still-pot, part of the liquid in the still-pot will vaporize and assuming that the components of the liquid mixture have a different vapour pressure, the vapour leaving the still-pot will be enriched in the components with a higher vapour pressure resulting in a distillate in the receiver different from the liquid mixture in the still pot. By evaporating part of the liquid in the still-pot the initial mixture in the still-pot is separated into two fractions, i.e. the residue in the still-pot and the distillate in the receiver. The separation effect will be small, however, unless the compositions of the liquid and the vapour produced from the liquid are substantially different.

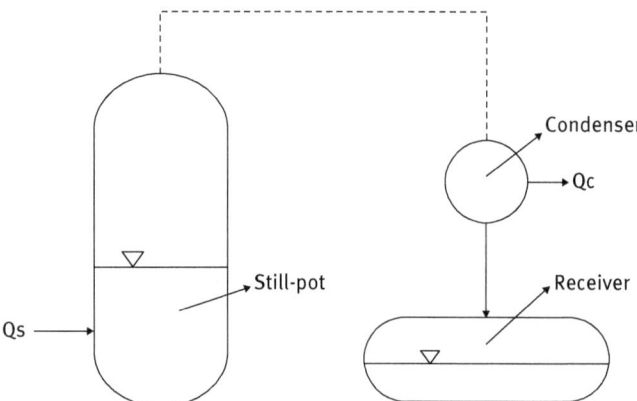

Fig. 1.1: Simple-Distillation.

https://doi.org/10.1515/9783110739732-001

1.2 Flash distillation

In a flash distillation according to Fig. 1.2 the mixture to be separated (Feed) is fed continuously to the flash-drum and with the addition of heat (Qs) separated into the two fractions distillate and bottom product. The amount of the two fractions depends on the amount of heat added and the separation effect is, like in simple distillation, given by the thermodynamic properties of the feed. Again, the separation effect will be small unless the vapour pressure of the components differs substantially.

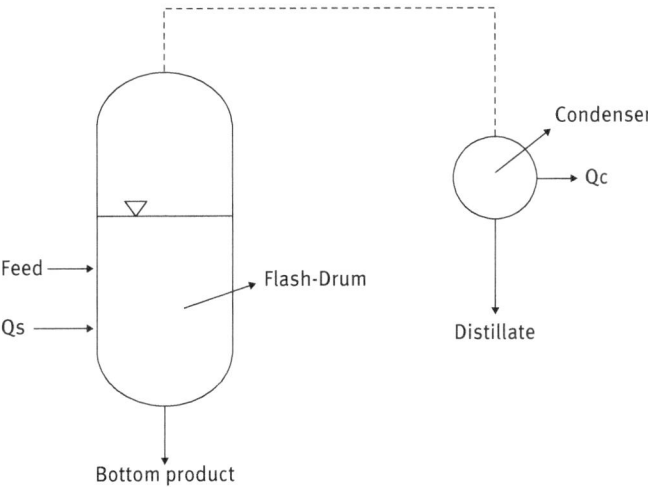

Fig. 1.2: Flash-Distillation.

1.3 Multistage distillation

In the simple as well as the flash distillation it is not possible to obtain one of the components as an almost pure component since the original mixture is separated into two fractions with all components present but at a composition different from the original mixture. Only by a connection in series of the simple distillation or the flash distillation in form of a cascade, the so-called rectification or multistage separation process as shown in Fig. 1.3, is it possible to separate a mixture into fractions with lesser components as the original mixture or even into almost pure components. Since most technical distillations are operated in the continuous mode, the batchwise distillation will not be discussed further.

In practice the cascade is realized in form of a tower or column with the addition of the feed at an optimized location and a counter current flow of the liquid and the vapour within the rectifying section above and the stripping section below the feed location. At the top of the column the distillate is withdrawn either as a vapour (not

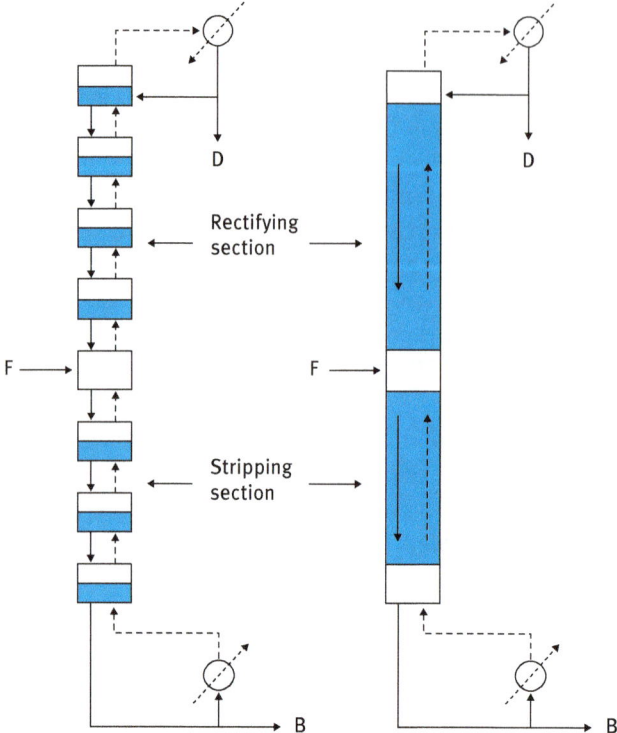

Fig. 1.3: Distillation cascade (– liquid flow - - vapour flow, F = Feed, D = Distillate, B = Bottom product).

shown in Fig. 1.3) or as part of the condensed vapour with the rest of the condensate returned to the column in order to provide for the liquid down-flow in the column. The liquid withdrawn from the bottom of the column is divided into the bottom product with the rest vaporized and returned to the column providing the necessary vapour up-flow. The internals of the column consist either of horizontal stages approximating the connection in series of the flash-drums of the cascade as shown by the column to the left in Fig. 1.3 or the column sections are filled with a packing like Raschig rings e.g. as illustrated by the column to the right in Fig. 1.3 [17]. The internals serve to increase the residence time of the liquid in the column, to provide for a large interfacial area between the liquid and the vapour flow and to enhance the mass transfer between the flows in the column.

Since the vapour rising from a boiling liquid is always enriched in the lower boiling components of the liquid, the composition of the lower boiling components of the vapour flowing upwards in the column increases from the bottom to the top of the column. Thus, starting from a given composition at the feed location, the composition of the vapour in the section above the feed location will be enriched in the lower boiling

components whereas the liquid flowing downwards in the section below the feed location will be stripped of the lower boiling components. The section of the column above the feed location is called the rectifying section, therefore, and the section below the feed location is the stripping section of a column.

Due to this separation effect the composition of the distillate and the bottom product differ from the composition of the feed. With a sufficient length of the respective section it is possible to obtain an almost pure component or to split the feed into two fractions with the components of one fraction not present in the other fraction. Taking for instance a feed with three components A, B, C and a column with a sufficient length of the rectifying and the stripping section, it is possible to produce the fractions A/BC, A/ABC; AB/BC, AB/ABC, ABC/C, ABC/BC, ABC/ABC and AB/C whereas a split AC/B is impossible. It follows that a separation of a mixture of n components into almost pure single components requires at least $(n-1)$ columns. As the possible splits increase exponentially with the number of components, determination of the optimal sequence taking into account constraints like minimum investment and operating costs can become a rather formidable optimisation problem [18].

2 Assumptions and problem reduction

1. Ideal mixtures are defined as mixtures with constant relative volatilities.
2. The flow rates of the liquid and the vapour in the rectifying and stripping section of a distillation column are considered constant.
3. The enthalpy of the liquid and the vapour are considered constant. This assumption does not affect the principle behaviour of a distillation process. The effect of strongly different heats of vaporisation of the pure components of a mixture on the flow rates can be accounted for by using caloric fractions rather than mole fractions [19, 36].
4. Except for the simple distillation only continuous, steady-state distillation modes at a countercurrent flow of the liquid and the vapour phase are discussed.
5. Any continuous distillation plant consists in principle of mass transfer sections with a countercurrent flow of liquid and vapour which are separated or supplemented by feed and product sections as shown in Fig. 2.1. The basic equations of distillation will be developed, therefore, on the basis of the distillation in a mass transfer section and subsequently applied to binary and multicomponent mixtures and the distillation in columns.
6. Except for the reversible distillation, all distillation columns operate under adiabatic conditions.

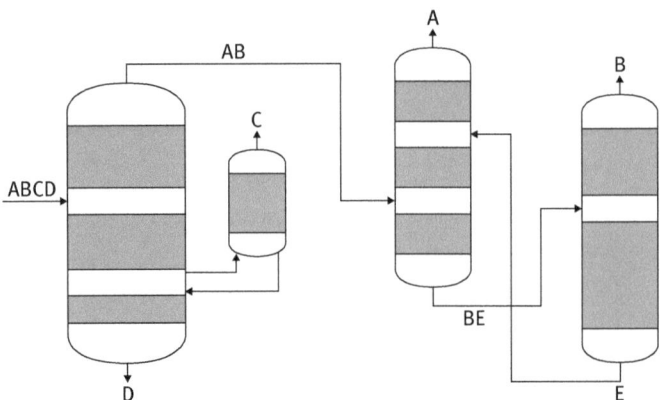

Fig. 2.1: Distillation plant (Filled areas = mass transfer sections, free areas = feed or product sections, ABCD = components in the feed, E = entrainer)

https://doi.org/10.1515/9783110739732-002

3 The basic equations of distillation

The basic equations will be developed without taking reference to a mixture with a specific number of components.

3.1 Mass balance

The mass balance is based on a differential element of a mass transfer section as shown in Fig. 3.1.

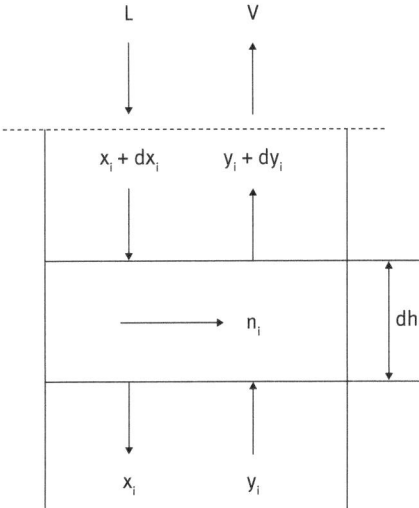

Fig. 3.1: Mass balance of a mass transfer section.

With the above assumption of constant flow rates of the vapour and the liquid, the component mass balance over the vapour space yields

$$V \cdot dy_i = n_i \cdot dA = n_i \cdot a \cdot A_S \cdot dh \tag{3.1}$$

and over the total differential section

$$V \cdot dy_i = L \cdot dx_i$$

or integrated

$$y_i = a_i + R \cdot x_i, \tag{3.2}$$

where a_i is a constant and R the flow ratio L/V.

Since equation (3.2) relates the mole fraction of a component in the vapour flow to the mole fraction of the respective component in the liquid flow at the different positions in the column it is called the operating line.

https://doi.org/10.1515/9783110739732-003

3.2 Vapour–liquid-equilibria

3.2.1 Ideal mixtures

The equilibrium of an ideal mixture is given by Dalton's law

$$p_i = y_i^* \cdot p \qquad (3.3)$$

in combination with Raoult's law

$$p_i = x_i \cdot p_i^0 \qquad (3.4)$$

resulting in

$$y_i^* \cdot p = x_i \cdot p_i^0. \qquad (3.5)$$

Taking into account

$$p = \sum x_j \cdot p_j^0 \qquad (3.6)$$

it follows

$$y_i^* = \frac{x_i \cdot p_i^0}{\sum x_j \cdot p_j^0}. \qquad (3.7)$$

Dividing the vapour pressures in equation (3.7) by a reference vapour pressure p_r^0 results in

$$y_i^* = \frac{\alpha_{ir} \cdot x_i}{E}, \qquad (3.8)$$

where

$$\alpha_{ir} = \frac{p_i^0}{p_r^0} \qquad (3.9)$$

is the so-called relative volatility and

$$E = \sum \alpha_{jr} \cdot x_j \qquad (3.10)$$

the mole averaged relative volatility.

The reference vapour pressure p_r^0 may be chosen arbitrarily. For reason of clarity, the vapour pressure of the highest boiling component is taken as the reference vapour pressure which – ordering the components in the order of an decreasing vapour pressure – gives the lowest boiling component the highest α-value and the highest boiling component an α-value of one.

3.2.2 Real mixtures

The vapour-liquid-equilibrium of real mixtures may be approximated to a high degree by treating the vapour phase as an ideal gas and by expanding Raoult's law (3.4) through a correction function, the so-called activity coefficient γ_i, i.e.

$$p_i = \gamma_i \cdot x_i \cdot p_i^0 \qquad (3.11)$$

resulting in

$$\alpha_{ir} = \frac{\gamma_i \cdot x_i \cdot p_i^0}{\gamma_r \cdot x_r \cdot p_r^0}. \tag{3.9a}$$

From the many correction functions available, here the Wilson equation in the form

$$\ln \gamma_i = -\ln \left(\sum_j x_j \cdot \Delta_{ij} \right) + 1 - \sum_l \frac{x_l \cdot \Delta_{li}}{\sum_j x_j \cdot \Delta_{lj}} \tag{3.12}$$

is used where the coefficients Δ_{ij}, Δ_{li} and Δ_{lj} are considered constant and are taken from the literature [20].

3.3 Mass transfer correlations

The most comprehensive description of the mass transfer from an ideal gas into another phase is the Maxwell–Stefan equation [21]

$$\nabla y_i = y_i \cdot \sum_j \frac{N_j}{\rho_G \cdot D_{ij}} - N_i \cdot \sum_j \frac{y_j}{\rho_G \cdot D_{ij}}. \tag{3.13}$$

If the binary diffusion coefficients D_{ij} are equal and if the total flux

$$N = \sum_j N_j \tag{3.14}$$

is zero, assumptions which are approximately valid for most mixtures in distillation, the Maxwell–Stefan equation reduces to Fick's equation

$$n_i = \rho_G \cdot D_{ij} \cdot \nabla y_i. \tag{3.15}$$

This equation may be solved on the basis of the film theory proposed by Whitman [22] resulting in the mass transfer equation

$$n_i = k_V \cdot (y_i^* - y_i). \tag{3.16}$$

This equation is strictly speaking a defining equation of the mass transfer coefficient k_V but has proven to allow for a sufficiently accurate solution of most technical mass transfer problems.

The above equations are also valid for the mass transfer of an ideal liquid by replacing the vapour variables by the respective liquid variables, i.e. the mass transfer equation for the liquid reads

$$n_i = k_L \cdot (x_i - x_i^*), \tag{3.17}$$

where x_i, y_i and x_i^*, y_i^* are the mole fractions in the bulk and at the interface of the liquid and the vapour, respectively, as shown schematically in Fig. 3.2.

Divison of equation (3.17) by (3.16) yields the ratio of the mass transfer coefficients

$$k = \frac{k_L}{k_V} = \frac{(y_i^* - y_i)}{(x_i - x_i^*)}. \tag{3.18}$$

For a given mass transfer coefficients ratio the mole fractions at the interface follow from

$$y_i^* = y_i + k \cdot (x_i - x_i^*) \tag{3.19}$$

as a function of the bulk mole fractions and from the assumption that the liquid and the vapour are in equilibrium at the vapour–liquid-interphase.

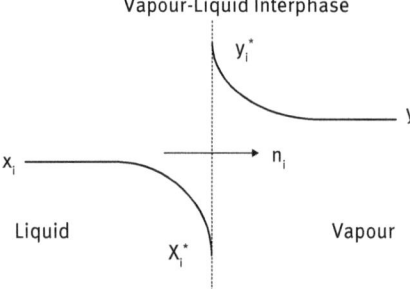

Fig. 3.2: Concentration profiles in two-phase mass transfer.

3.4 Distillation in a packed mass transfer section

Combining the mass balance equation (3.1) with the mass transfer correlation (3.16) or (3.17) gives

$$n_i \cdot a \cdot A_S \cdot dh = V \cdot dy_i = k_V \cdot (y_i^* - y_i) \cdot a \cdot A_S \cdot dh \tag{3.20}$$

and

$$n_i \cdot a \cdot A_S \cdot dh = L \cdot dx_i = k_L \cdot (x_i - x_i^*) \cdot a \cdot A_S \cdot dh, \tag{3.21}$$

respectively.

3.4.1 Distillation at total reflux ($L/V = 1$)

If the flows of liquid and vapour are equal (total reflux), then $y_i = x_i$ and equations (3.20) and (3.21) can be rearranged as

$$\frac{dx_i}{y_i^* - x_i} = \frac{k_V \cdot a \cdot A_S}{V} \cdot dh = \frac{dh}{HTU_V} = d(NTU_V) \tag{3.22}$$

and

$$\frac{dx_i}{x_i - x_i^*} = \frac{k_L \cdot a \cdot A_S}{L} \cdot dh = \frac{dh}{HTU_L} = d(NTU_L) \tag{3.23}$$

with NTU as the number of transfer units and HTU as the height of a transfer unit [23].

3.4.2 Effect of the ratio of the mass transfer coefficients

3.4.2.1 Mass transfer resistance exclusively in the vapour phase ($k = \infty$)
Rearranging equation (3.22) yields

$$\frac{dx_i/x_i}{d(NTU)_v} + 1 = \frac{d\ln x_i + d(NTU)_V}{d(NTU)_V} = \frac{y_i^*}{x_i} \tag{3.22a}$$

and for component k

$$\frac{dx_k/x_k}{d(NTU)_v} + 1 = \frac{d\ln x_k + d(NTU)_V}{d(NTU)_V} = \frac{y_k^*}{x_k}, \tag{3.22b}$$

Division of the two equations results in

$$\frac{d\ln x_i + d(NTU)_V}{d\ln x_k + d(NTU)_V} = \frac{y_i^* \cdot x_k}{x_i \cdot y_k^*} = \alpha_{ik} \tag{3.24}$$

and integration yields

$$NTU_V = -\ln\frac{x_i}{x_{i0}} + \frac{1}{1-\alpha_{ki}}\left(\ln\frac{x_i}{x_{i0}} - \ln\frac{x_k}{x_{k0}}\right) = \frac{H_V}{HTU_V} = H_V^*. \tag{3.25}$$

3.4.2.2 Mass transfer resistance exclusively in the liquid phase ($k = 0$)
If the mass transfer resistance is located in the liquid phase only, the mole fraction of the liquid at the interface x_i^* is in equilibrium with the mole fraction of the vapour in the bulk of the vapour y_i which – due to the assumption of equal flow rates of vapour and liquid – is equal to x_i. The relation is the mirroring of the equilibrium curve equation (3.8) about the line $f(x) = x$, i.e. the relative volatility α_{ik} in equation (3.8) has to be replaced by the reciprocal value

$$\frac{1}{\alpha_{ik}} = \alpha_{ki} \tag{3.26}$$

resulting in [24]

$$x_i^* = \frac{1/\alpha_{ik} \cdot x_i}{\sum(1/\alpha_{jk} \cdot x_j)} = \frac{\alpha_{ki} \cdot x_i}{\sum(\alpha_{kj} \cdot x_j)}. \tag{3.27}$$

Replacing x_i^* in equation (3.22) by this equation and integration yields

$$NTU_L = -\ln\frac{x_i}{x_{i0}} + \frac{1}{1-\alpha_{ik}} \cdot \left(\ln\frac{x_i}{x_{i0}} - \ln\frac{x_k}{x_{k0}}\right) = \frac{H_L}{(HTU)_L} = H_L^*. \tag{3.28}$$

It should be noted that equations (3.25) and (3.28) apply to any combination of two components of a mixture.

3.5 Distillation in a mass transfer section with theoretical stages

A distillation column consisting of a cascade of flash-drums as illustrated by the left column in Fig. 1.3 is called a column with theoretical stages in which a theoretical stage is defined as a black box where the outgoing flows are in thermodynamic equilibrium [3, 4] as shown in Fig. 3.3. This definition has the advantage that the concentration profile can be calculated exclusively on the basis of the mass balance and the equilibrium relation.

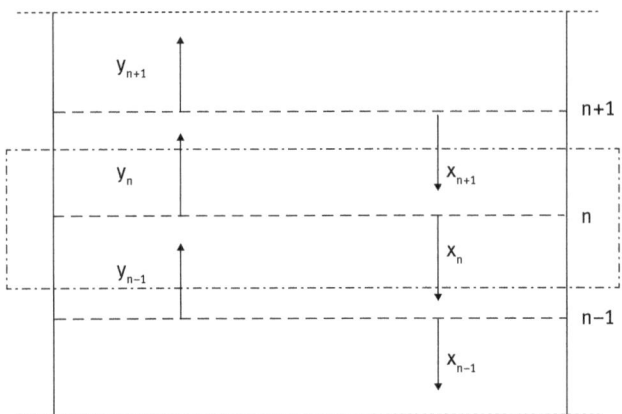

Fig. 3.3: Theoretical Stage Concept (- - - theoretical stage, · · · mass balance).

The "Theoretical Stage Concept" at total reflux $L/V = 1$ is the forward difference solution of the differential equation (3.22) with

$$\left.\frac{dx}{d(NTU)_{TS}}\right|_n = \frac{x_{n+1} - x_n}{\Delta((NTU)_{TS})} = y_n - y_{n-1}. \tag{3.22c}$$

Setting

$$\Delta((NTU)_{TS}) = \text{one theoretical stage}$$

yields with the mass balance in Fig. 3.3

$$y_n - y_{n-1} = x_{n+1} - x_n \,,$$

which is equivalent to equation (3.18) and indicates that the ratio of the mass transfer resistance in the liquid and vapour phase is equal to one at total reflux as illustrated in Fig. 3.4.

For the total reflux case and a constant relative volatility Fenske [7] and Underwood [8] developed the analytical solution

$$NTS = \frac{\ln\left(\frac{x_i}{x_{0,i}} \cdot \frac{x_{0,k}}{x_k}\right)}{\ln \alpha_{ik}} = \frac{\ln\left(x_i/x_k\right) - c_{ik}}{\ln \alpha_{ik}} \,, \tag{3.29}$$

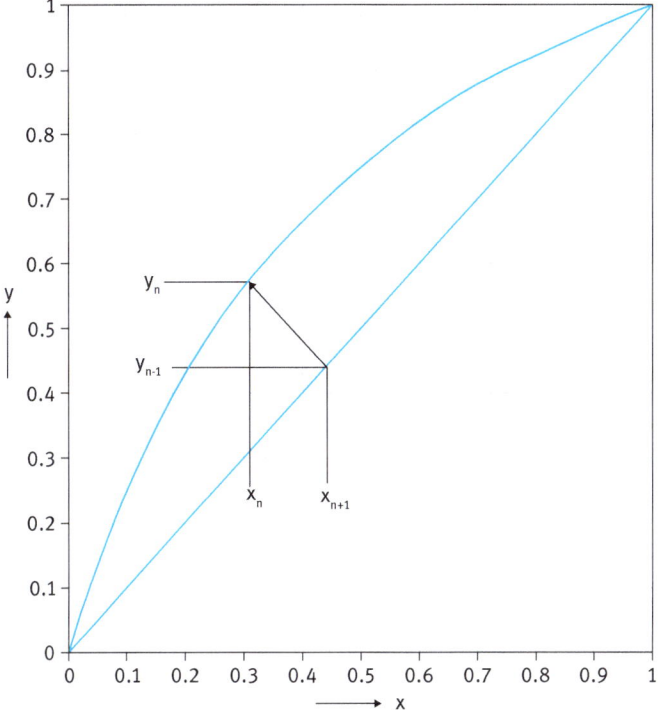

Fig. 3.4: McCabe–Thiele-diagram illustrating the Theoretical Stage Concept.

where NTS is the number of theoretical stages required to achieve the desired separation.

As will be shown in the context of ternary mixtures the equation (3.29) corresponds to an equal distribution of the mass transfer resistance in both phases.

3.6 Distillation at partial reflux ($L \neq V$)

The development of a generally valid equation will be based on a binary mixture as shown in Fig. 3.5 for better visualization. The results are valid for any number of components, however, as will be shown later.

The principal difference between a distillation at a flow ratio of $L/V \neq 1$ and a distillation at total reflux with a flow ratio $L/V = 1$ is that the operating line according to equation (3.2) does not intersect with the equilibrium curve according to equation (3.8) at the positions

$$y_i^* = y_i = x_i = 0 \quad \text{and} \quad y_i^* = y_i = x_i = 1 \tag{3.30}$$

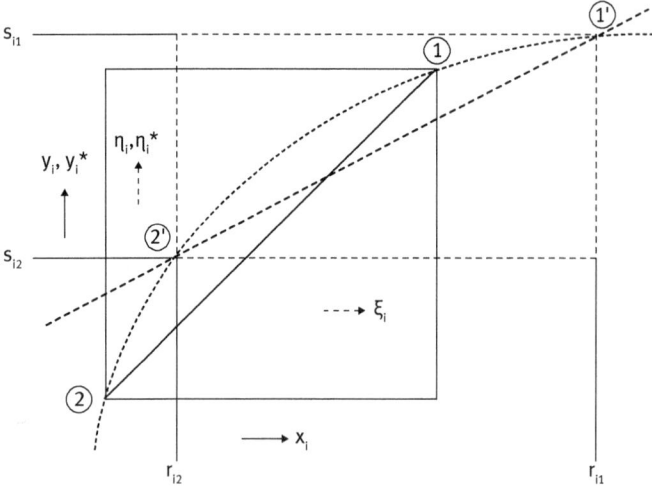

Fig. 3.5: Binary distillation (dotted line = equilibrium line, full lines = binary system, dashed lines = transformed binary system).

of the coordinate system

$$y_i^*, y_i = f(x_i) \tag{3.31}$$

but rather at the positions

$$y_i^* = y_i = a_i + R \cdot x_i \,, \tag{3.32}$$

i.e. the distillation endpoints defined by the intersections of the operating line with the equilibrium curve are shifted and may even be located outside the physically feasible domains $0 \leq x, y \leq 1$.

In order to obtain the same conditions for $R = L/V \neq 1$ as for $R = L/V = 1$ a new coordinate system is introduced [12]

$$\eta_i^*, \eta_i = f(\xi_i) \,, \tag{3.33}$$

which fulfills the condition

$$\eta_i^* = \eta_i = \xi_i = 0 \quad \text{and} \quad \eta_i^* = \eta_i = \xi_i = 1 \tag{3.34}$$

at the intersection of the equilibrium line and the balance line as well as

$$\eta_i = \xi_i \tag{3.35}$$

for all values of ξ_i.

This is achieved by a linear transformation between the y_i^*, y_i, x_i- and the η_i^*, η_i, ξ_i-system. For this purpose, the mole fractions at the intersections of the equilibrium line with the balance line (see point 1 and 2 in Fig. 3.5) are defined by the variable r_{in}

for the x_i-coordinate and s_{in} for the y_i^*, y_i-coordinate, where the index i stands for the component i and the index n for the different intersections.

From a physical point of view the intersections $y_i^* = y_i$ mark the limiting concentrations that can be obtained in a column section of infinite length and are called a "node".

It should be noted that a node is uniquely defined by the equilibrium equation and the mass balance equation and does not depend on the mass transfer model used.

If we denote the nodes in the same way as the sequence of the components, i.e. the right node in Fig. 3.5 with $n = 1$ and the left node with $n = 2$, it follows

$$\xi_i = \frac{x_i - r_{i,2}}{r_{i,1} - r_{i,2}} \tag{3.36}$$

and

$$\eta_i = \frac{y_i - s_{i,2}}{s_{i,1} - s_{i,2}} \tag{3.37}$$

or for a mixture with n components in matrix notation

$$|x| = |r| \cdot |\xi| \tag{3.38}$$

and

$$|y| = |s| \cdot |\eta| \,, \tag{3.39}$$

where $|r|$ and $|s|$ are quadratic transformation matrices with the elements r_{in} and s_{in}, respectively.

The matrix elements r_{in} are obtained from equation (3.32) by replacing y_i^* by equation (3.8) and x_i by r_{in} resulting in

$$r_{in} = \frac{a_i \cdot E_n}{\alpha_i - R \cdot E_n} \tag{3.40}$$

and the matrix elements s_{in} from the balance equation

$$s_{in} = a_i + R \cdot r_{in} \,. \tag{3.41}$$

Since

$$\sum a_i \cdot r_{in} = E_n \tag{3.42}$$

it follows for any node n

$$\sum \frac{a_i \cdot E_n}{\alpha_i - R \cdot E_n} = 1 \,. \tag{3.43}$$

In the mathematical literature [25] equation (3.43) is called the characteristic equation or eigenfunction of the corresponding differential equation, the n roots of equation (3.43) for a n component system are the characteristic roots or the eigenvalues E_n and the coordinates η_i, ξ_i are the Eigencoordinates of the new system.

The eigenvalues E_n of the η_i^*, η_i, ξ_i-system correspond to the α_i-values in the y_i^*, y_i, x_i-system (see Appendix A.1), i.e.

$$\eta_i^* = \frac{E_i \cdot \xi_i}{E} \,, \tag{3.44}$$

where

$$E = \sum_i E_i \cdot \xi_i. \tag{3.45}$$

All of the above in Chapter 3 developed equations thus also apply to the distillation at any flow rate of liquid and vapour by replacing (see Appendix A.1).

$$E_i \to \alpha_i \tag{3.46}$$

$$\eta_i^* \to y_i^* \tag{3.47}$$

$$\eta_i \to y_i \tag{3.48}$$

$$\xi_i \to x_i \tag{3.49}$$

The numerical values in the η_i, ξ_i-space are obtained by the inverse of the equations (3.36) to (3.39).

3.7 Simple distillation

The theoretical solution of the composition of the liquid in the still-pot during simple distillation according to Fig. 1.1 was given by Ostwald [26] and Rayleigh [6] based on the unsteady state differential mass balance equations at a constant vapour rate

$$d(x_i \cdot L_S) = L_S \cdot dx_i + x_i \cdot dL_S = -y_i^* \cdot V \cdot dt \tag{3.50}$$

$$dL_S = -V \cdot dt \tag{3.51}$$

or

$$\frac{dL_S}{L_S} = d \ln L_S = \frac{dx_i}{y_i^* - x_i}. \tag{3.52}$$

Because of the analogue structure of the equations (3.22) and (3.52) it follows from equation (3.26)

$$\frac{L_S}{L_{S,0}} = \frac{x_{i0}}{x_i} \cdot \left(\frac{x_i \cdot x_{k0}}{x_{i0} \cdot x_k} \right)^{\frac{\alpha_{ik}}{\alpha_{ik}-1}}, \tag{3.53}$$

i.e. the concentration profile of the liquid in the still-pot of a simple distillation, which is called the residue curve, is identical to the concentration profile of the liquid in a mass transfer section at an equal flow rate of liquid and vapour and a mass transfer resistance totally on the gas side with the difference that NTU_V is replaced by L/L_0.

Equation (3.53) is also the solution of the continuous falling-film distillation if L is taken as the flow rate of the liquid with the other assumptions analogue to those in simple distillation. The difference of these modes of distillation is the frame of reference, i.e. simple distillation is defined in the LaGrange frame of reference whereas the falling-film distillation is defined in the Euler frame of reference [27].

Even though in many papers residue curves are used to describe the concentration profiles in continuous distillation, here the concentration profiles in continuous distillation will be called distillation lines in order to avoid any unnecessary confusion.

3.8 Reversible distillation

In reversible distillation the concentration of the vapour at any location in the mass transfer section is in equilibrium with the concentration of the liquid phase at this location, i.e. the operating line of the column coincides with the vapour–liquid-equilibrium curve [28].

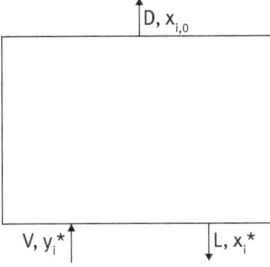

D, $x_{i,0}$

V, y_i^* L, x_i^* **Fig. 3.6:** Mass balance of a reversible distillation.

The concentration profile of a reversible distillation follows from the mass balances (Fig. 3.6)

$$V = D + L, \tag{3.54}$$

$$V \cdot y_i^* = D \cdot x_{i,0} + L \cdot x_i^* \tag{3.55}$$

and the equilibrium equation (3.8). Elimination of D and rearrangement yields

$$\alpha_i = E \cdot \left[(1 - R) \cdot \left(\frac{x_{i,0}}{x_i} - 1 \right) + 1 \right]. \tag{3.56}$$

For a binary system equation (3.56) only states that the mole fraction of the vapour at any location in the column is in equilibrium with the mole fraction of the liquid at the same location which follows already from the definition of the reversible distillation.

Reversible distillation uses the minimum energy possible for a desired separation and thus serves as a benchmark for comparing different modes of distillation. In addition, it is useful in obtaining the optimum separation sequence with the minimal energy requirements in the distillation of multicomponent mixtures as will be discussed later.

4 Distillation of ideal mixtures

4.1 Binary mixtures

4.1.1 Distillation at total reflux

4.1.1.1 Mass transfer section with a packing

The relevant equations to be applied are the equations (3.25) and (3.28) with the indices $i = 1$, $k = 2$ and $x_k = 1 - x_i$. The concentration profiles are given in Fig. 4.1 for a concentration change $0.01 < x_1 < 0.99$ as a function of the number of transfer units with the mass transfer resistance on the vapour side ($k = \infty$), an equal distribution of the mass transfer coefficients ($k = 1$) and for a mass transfer resistance completely on the liquid side ($k = 0$).

It is worth noticing that the curve with the mass transfer completely on the vapour side ($k = \infty$) and the curve with the resistance completely on the liquid side ($k = 0$) show a stronger concentration change related to one transfer unit in the top region and the bottom region of the distillation column, respectively.

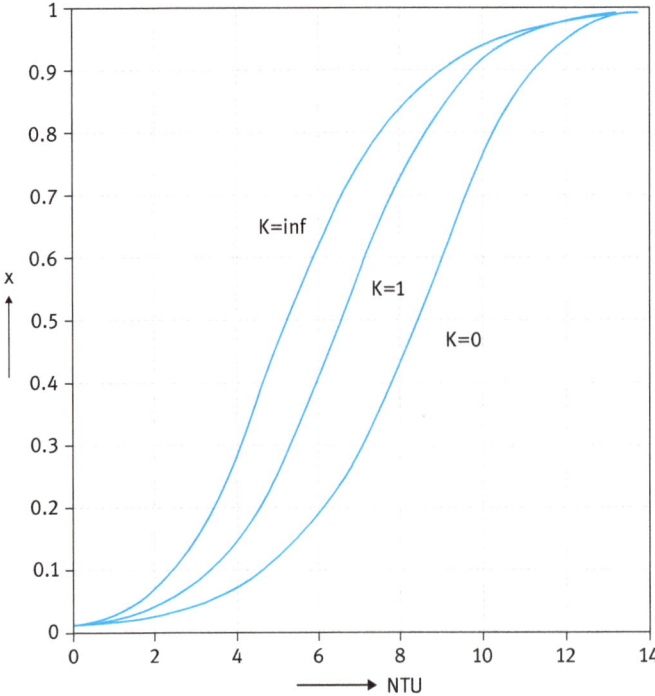

Fig. 4.1: Binary concentration profiles vs. the number of theoretical stages (total reflux, $\alpha = [2\ 1]$).

https://doi.org/10.1515/9783110739732-004

4.1.1.2 Mass transfer section with theoretical stages

The relevant equation to be applied is the equation (3.29) with the index $i = 1$, index $k = 2$ and $x_2 = 1 - x_1$. The concentration profiles are given in Fig. 4.2 for a concentration change $0.001 < x_1 < 0.999$ as a function of the number of theoretical stages.

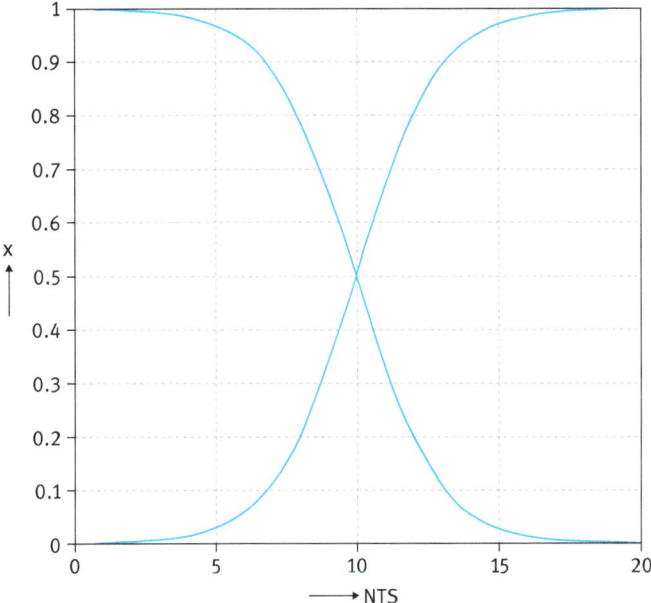

Fig. 4.2: Binary concentration profiles vs. the number of theoretical stages.

In contrast to the concentration profiles in packed mass transfer sections which depend on the mass transfer resistance in the liquid and vapour phase, the concentration profiles in a mass transfer section with theoretical stages are symmetrical with respect to the height of the column. This is due to the fact that the concept of a theoretical stage implies that the mass transfer resistance in the liquid and the vapour are equal [14, 29].

4.1.2 Distillation at partial reflux

4.1.2.1 Mass transfer section with a packing or theoretical stages

The principal procedure to be followed in the general case $R = L/V \neq 1$ will be explained taking the rectifying section of a distillation as an example. The calculations are based on (see Chapter 3.6):

$$\alpha_1 = 2, \alpha_2 = 1, x_F = 0.5, x_{1,D} = 0.9, R = 0.78, a_1 = 0.2$$

resulting in Fig. 4.3.

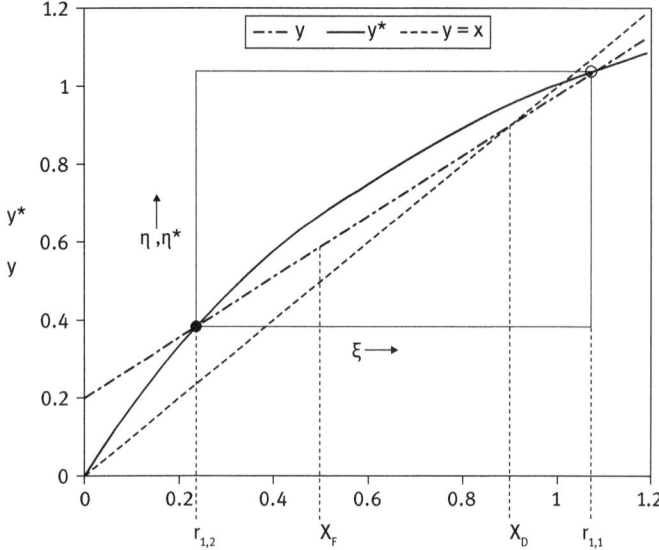

Fig. 4.3: Mass transfer section with different flow rates of liquid and vapour (solid line = equilibrium curve, dashed line = diagonal of the diagram, dashed-dotted line = operating line with a slope $R < 1$).

The eigenvalues follow from equation (3.43) with $E_1 = 2.070$, $E_2 = 1.240$ and the x_1-coordinates of the intersections of the operating line with the equilibrium curve from equation (3.40)

$$r_{12} = 0.240, r_{11} = 1.074.$$

For calculating the dimensionless H^*-values or the number of theoretical stages required for a change in the mole fraction from $x_{1,F} = 0.5$ to $x_{1,D} = 0.9$ a transformation of these boundary mole fractions into the ξ-coordinates applying equation (3.36) yields

$$\xi_{10} = 0.312, \xi_{20} = 0.688 \quad \text{and} \quad \xi_1 = 0.791, \xi_2 = 0.209$$

and by replacing the x_i, x_j and the α_i, α_j in the equations (3.26), (3.28) and (3.29) by the respective ξ_i, ξ_j and E_i, E_j it follows

$$H_V^* = 4.35, \quad H_L^* = 4.08 \quad \text{and} \quad NTS = 4.14.$$

It should be noted that a transformation of the mole fractions of the vapour is not necessary as these are not required in calculating the dimensionless heights or number of theoretical stages. The transformation of the stripping section follows the same procedure as outlined above for the rectifying section as will be shown later.

4.1.3 Distillation column with a rectifying and a stripping section

The basic distillation column consists of two mass transfer sections with a rectifying and a stripping section separated by the feed section, a partial or total condenser at the top and a partial or total reboiler at the bottom of the column. For sake of simplicity it is assumed in the following that the columns are equipped with a total condenser and a total reboiler. Such a column has $(c + 6)$ degrees of freedom [17]. These degrees of freedom can be utilized e.g. by the following specifications

– the complete definition of the feed	$c + 2$
– the operating pressure of the column	1
– the mole fraction of one component in the distillate	1
– the mole fraction of one component in the bottom product	1
– the feed location	1

Based on these specifications, the required height or the number of theoretical stages of the mass transfer sections can be calculated.

It is important to note that any degree of freedom listed in the above list can be replaced by another independent variable, i.e. the mole fraction of one component in the bottom product can be replaced e.g. by the flow rate of the distillate.

4.1.3.1 McCabe–Thiele-diagram

The McCabe–Thiele-diagram as shown in Fig. 4.4 allows for a graphical solution of a binary distillation problem at constant flow ratios in the column [30]. The minimum flow ratio is given by the intersection of the operating line with the q-line as discussed above unless the equilibrium curve has an inflection point. In such a case the limiting flow ratio is given by the operating line which is a tangent to the equilibrium curve.

The number of theoretical stages of a staged column with specified feed and product compositions at any given flow ratio may be obtained by a graphical iteration in the McCabe–Thiele-diagram. Drawing a vertical line through the given bottom product x_B, this line will intersect with the equilibrium curve and according to the definition of a theoretical stage yield the composition of the vapour y^* leaving the lowest theoretical stage. Next a horizontal line passing through y^* will intersect with the operating line $y = f(x)$ and give the composition of the liquid flow leaving the next higher stage which in turn yields the equilibrium vapour rising from the next higher stage (see Fig. 4.4). This procedure will be continued until the horizontal line has intersected with the q-line indicating that the rectifying section of the column has been entered and the iteration has to proceed with the operating line of the rectifying section until the specified distillate composition has been overstepped.

For a packed column with specified feed and product compositions the required number of transfer units for any flow ratio $R_{min} < R \leq 1$ is calculated applying equation (3.22) or (3.23) and the equation (3.2) of the operating line with the driving forces

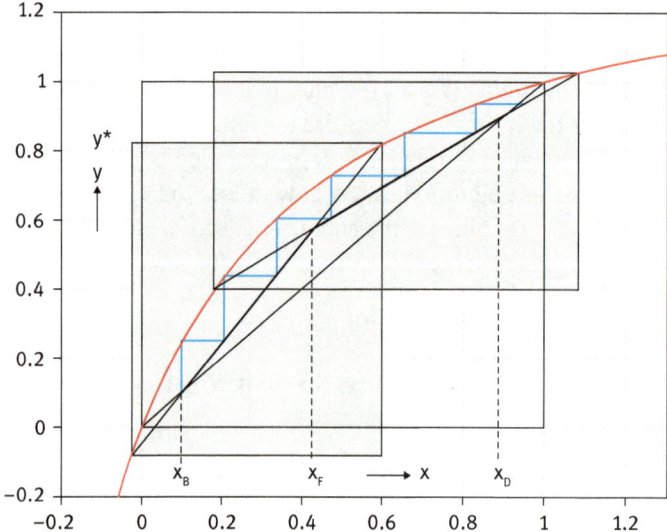

Fig. 4.4: McCabe–Thiele-diagram (The left shaded rectangle is the distillation space of the stripping section and the right rectangle the distillation space of the rectifying section.).

taken from the McCabe–Thiele-diagram. The height of the column is given by the product of $NTU \cdot HTU$.

The minimum number of stages or transfer units is obtained at total reflux ($R = S = 1$) and at the minimum flow ratio of the rectifying section and the related maximum flow ratio of the stripping section the number of stages or transfer units becomes infinite. The optimum flow ratio follows from an economic optimisation.

The number of theoretical stages or the number of transfer units may be calculated taking into account that by a coordinate transformation, as discussed above in the sections 3.5 and 3.6, the rectifying and the stripping section are changed to systems with a flow ratio of one as shown by the rectangles in Fig. 4.4. The rectangles are defined by the intersections of the operating lines of the two sections with the equilibrium curve and the respective relative volatilities of the two systems are given by the E-values of these intersections, i.e. $E_R = \alpha_R = [3.16\ 1.36]$ and $E_S = \alpha_S = [2.20\ 0.96]$, respectively. The number of theoretical stages or transfer units of the two sections then follow from equation (3.29) and (3.25, 3.28) based on the transformed product coordinates of the distillation spaces with the results: $NTS_R = 2.76$, $NTU_R = 3.09$ and $NTS_S = 2.71$, $NTU_S = 2.81$, respectively.

4.1.3.2 Limiting flow ratios (Minimum reflux condition)

The condition of limiting flow ratios is given if the operating lines of the rectifying and the stripping section intersect with the equilibrium curve as shown for example in Fig. 4.5. The limiting condition is an important quantity in distillation as it defines

the minimum and the maximum possible slope of the operating line of the rectifying and the stripping section of a distillation task, respectively. Whereas the number of transfer units or the number of theoretical stages becomes infinite both in the rectifying and the stripping section of the column since the driving forces $|y^* - y|$ are zero at this intersection, the required energy of separation is at its minimum.

Although from a mass transfer point of view the flow ratios R in the rectifying and S in the stripping section of a column are the characteristic variables of a mass transfer section, it is common practice in the literature to apply the reflux ratio

$$r = L/D = R/(1 - R) \tag{4.1a}$$

defined as the flow of the liquid returned to the top of the column divided by the flow rate of the distillate for the rectifying and the reboil ratio

$$s = V/B = 1/(S - 1) \tag{4.1b}$$

defined as the flow of the vapour returned to the bottom of the column divided by the flow rate of the bottom product for the stripping section as the more relevant variables from a column operation point of view.

The minimum flow ratio of the rectifying section is equal to the minimum possible slope of the operating line in the rectifying section

$$R_{\min} = \frac{(y_D - y_F^*)}{(x_D - x_F)} \tag{4.2a}$$

and the maximum flow ratio of the stripping section equal to the maximum possible slope of the operating line in the stripping section

$$S_{\max} = \frac{(y_B - y_F^*)}{(x_B - x_F)}, \tag{4.2b}$$

where y_F^* and x_F are the coordinates of the intersection of the operating lines with the vapour-liquid-equilibrium curve (see Fig. 4.5).

The minimum flow ratios may also be calculated by applying the characteristic function of the rectifying and stripping section. Replacing the a_i in equation (3.43) by the respective product composition

$$a_i = (1 - R) \cdot x_{D,i} \qquad \text{and} \tag{4.3a}$$
$$a_i = (1 - S) \cdot x_{B,i}, \tag{4.3b}$$

where S is the flow ratio of the stripping section, and multiplication by the related product flow rates yields

$$\frac{1}{(1 - R)} = \sum_i \frac{a_i \cdot x_{D,i}}{a_i - R \cdot E_{R,n}} \qquad \text{and} \tag{4.4a}$$

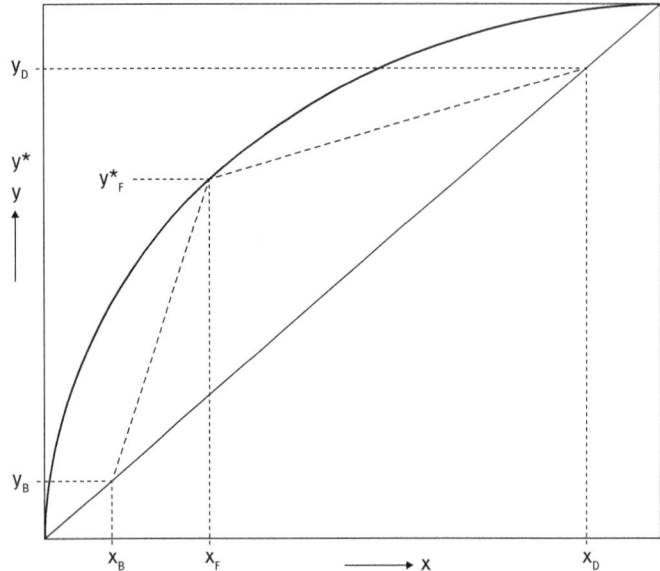

Fig. 4.5: Limiting flow ratios (curved line = equilibrium curve, dashed lines = operating lines of the rectifying and stripping section at minimum flow ratio R_{min} and maximum flow ratio S_{max}).

$$\frac{1}{(1 - S)} = -\sum_i \frac{\alpha_i \cdot x_{B,i}}{\alpha_i - S \cdot E_{S,m}} \cdot \quad (4.4b)$$

Inserting the respective product flow rates into the two equations and addition gives

$$\frac{D}{(1 - R)} + \frac{B}{(1 - S)} = \sum_i \frac{\alpha_i \cdot D \cdot x_{D,i}}{\alpha_i - R \cdot E_{R,n}} - \sum_i \frac{\alpha_i \cdot B \cdot x_{B,i}}{\alpha_i - S \cdot E_{S,m}} \cdot \quad (4.5)$$

Since the left side of equation (4.5) can be written as

$$\frac{D}{(1 - R)} + \frac{B}{(1 - S)} = (V_R - V_S) = (1 - q) \cdot F \quad (4.6)$$

and taking into account the balance equation

$$D \cdot x_{D,i} + B \cdot x_{B,i} = F \cdot x_{F,i}, \quad (4.7)$$

there must be at least one common product

$$R \cdot E_{R,n} = S \cdot E_{S,m} = \Phi \quad (4.8)$$

reducing equation (4.5) to

$$1 - q = \sum_i \frac{\alpha_i \cdot x_{F,i}}{\alpha_i - \Phi} \cdot \quad (4.9)$$

For a mixture with n components and flow ratios $L/V \neq 1$ there are always $(n - 1)$ roots Φ.

Inserting the root $\Phi = \Phi_{key}$ with a numerical value between the relative volatilities of the light and the heavy key component into equations (4.4a) and (4.4b) yields the minimum flow ratio of the rectifying section

$$\frac{1}{1 - R_{min}} = \sum_i \frac{\alpha_i \cdot x_{D,i}}{\alpha_i - \Phi_{key}} \tag{4.10a}$$

and the maximum flow ratio of the stripping section

$$\frac{1}{1 - S_{max}} = -\sum_i \frac{\alpha_i \cdot x_{B,i}}{\alpha_i - \Phi_{key}}, \quad \text{respectively.} \tag{4.10b}$$

The flow rate S_{max} can also be calculated using the mass balance

$$S_{max} = (R_{min} \cdot D + (1 - R_{min}) \cdot q \cdot F)/(D - (1 - R_{min}) \cdot (1 - q) \cdot F) \tag{4.10c}$$

The same results were obtained by Underwood [9] for columns with theoretical stages as the location of the nodes of the distillation spaces does not depend on the model used for calculating the distillation lines.

The $(\Phi = \Phi_{key})$-value of equation (4.9) inserted into equations (4.10a) and (4.10b) yields the limiting flow ratios and from equation (4.8) then follows the E-value of one node of each of the two distillation sections. The missing E-value for the second node of the two sections is the E-value of the intersection of the two operating lines with the equilibrium curve E_I common to both distillation spaces. The coordinates of the nodes follow from equations (3.40), (3.41).

For the distillation presented in Fig. 4.6 the relevant data are: $\Phi_{key} = 1.667$, $R_{min} = 0.515$, $S_{max} = 1.762$, $E_I = 1.544$, $E_R = [3.235 \ 1.800]$, $E_S = [1.800 \ 0.946]$.

4.1.3.3 Optimal feed location

The optimal feed location for given flow ratios follows from the requirement of a minimum height of the column which is fulfilled if the driving force at any position in the column is equal to the maximum possible driving force at that position [17] and the condition of a minimal mixing of the feed with the flows at the feed location. Thus, if the feed consists of a liquid flow F_L with the mole fraction $x_{L,i}$ and a vapour flow F_V with the mole fraction $y_{V,i}$ then according to the mass balance

$$F_L \cdot x_{L,i} + F_V \cdot y_{V,i} = F \cdot x_{F,i} \tag{4.11}$$

the liquid and the vapour feed must be introduced into the column at a position where the mole fraction x_i in the liquid and the mole fraction y_i in the vapour are identical to the respective mole fractions in the feed, i.e. $x_i = x_{L,i}$ and $y_i = y_{V,i}$. Taking this into consideration equation (4.11) yields the so-called q-line

$$y_i = \frac{1}{1 - q} \cdot x_{F,i} - \frac{q}{1 - q} \cdot x_i, \tag{4.12a}$$

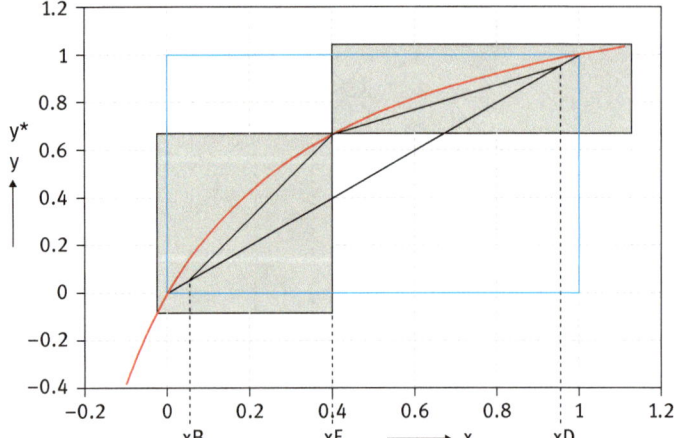

Fig. 4.6: Distillation spaces at limiting flow ratios (curved line = equilibrium curve with α = [3 1], full lines = limiting operating lines with x_B = 0.05, x_F = 0.4, x_D = 0.95, q = 1).

where

$$q = \frac{F_L}{F} \quad \text{and} \quad (1 - q) = \frac{F_V}{F} \tag{4.13}$$

or

$$q = \frac{y_i - x_{F,i}}{y_i - x_i}. \tag{4.12b}$$

Equation (4.13) defines the thermal state of the feed:

$q < 0$ the feed is an overheated vapour,
$q = 0$ the feed is a saturated vapour,
$1 < q < 0$ the feed is a two-phase feed,
$q = 1$ the feed is a saturated liquid,
$q > 1$ the feed is a subcooled liquid.

Equation (4.12a) is a straight line passing through the intersection of x_F with the diagonal as shown in Fig. 4.7.

At optimal conditions the operating lines of the two distillation sections intersect at the q-line as shown in Fig. 4.7 for a two-phase feed with q = 0.5. Beginning at the given bottom composition about 5.5 theoretical stages are required to reach the desired distillate composition with the flow ratios as given in Fig. 4.7.

4.1.3.4 Optimal flow ratios

The optimal flow ratios follow from an economic analysis minimizing the total costs of the distillation process consisting of the investment and operating costs as a function of the flow ratio of the rectifying section [17]. The analysis is based on the fact that the investment costs of a column are infinite at the limiting flow ratios due to the infinite

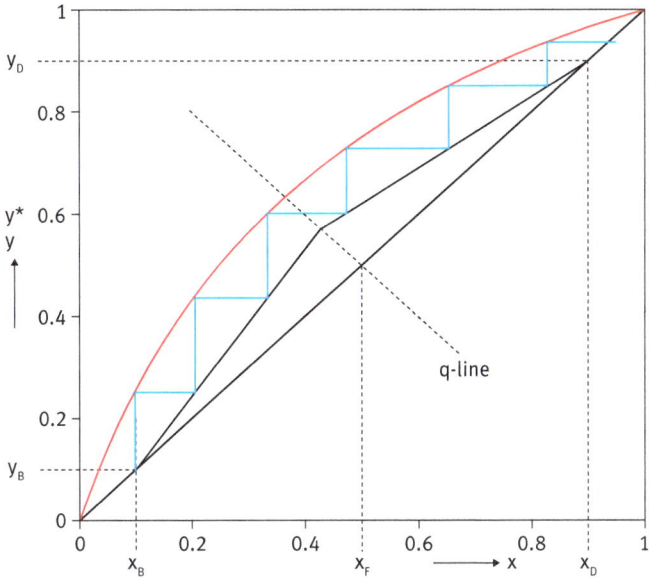

Fig. 4.7: Feed stage location in the McCabe–Thiele-diagram (curved line = equilibrium line, dashed lines = operating lines of the rectifying and stripping section, dash-dotted line = q-line).

number of stages or number of transfer units required and decrease with an increasing flow ratio of the rectifying section whereas the operating costs are at a minimum at the limiting flow ratios and increase with an increasing flow ratio of the rectifying section. Thus plotting the investment and the operating costs as a function of the flow ratio of the rectifying section gives two curves with opposite slopes resulting in a minimum of the total costs yielding the optimal flow ratio.

Whereas the height of a distillation column is determined by the mass transfer, the diameter of a column is given by the maximum acceptable flow rates, i.e. the fluid dynamics of the column.

The number of practical stages can be estimated with the so called overall Murphree efficiency E_M [17, 29, 31] of a distillation column defined as the ratio of the measured concentration difference between the top and the bottom of a column divided by the number of theoretical stages which has a value of about $E_M = 0.7$. Multiplying the number of practical stages by the spacing of the stages yields the approximate height of the column. For a column with a packing an average height of a transfer unit of $HTU = 0.4$ m can be used as a first estimate.

The fluid dynamics of a column are limited by the so-called flooding of the column due to an excessive rate of the vapour or liquid flow [17].

4.2 Ternary mixtures

Most discussions on ternary distillation use the concept of theoretical stages [9, 10, 32–34] even though the differential equation of distillation (3.20) as based on the physically founded mass transfer concept represents a more fundamental concept to which all other concepts are related. Writing equation (3.20) for the components i and j and assuming that the mass transfer coefficient is identical for both components and dividing the two equations yields the differential equation of the concentration profiles in ternary distillation, the so-called distillation lines

$$\frac{dx_j}{dx_i} = \frac{(y_j^* - y_j)}{(y_i^* - y_i)} . \tag{4.14}$$

4.2.1 Distillation at total reflux and simple distillation

At total reflux the flow ratio L/V of the mass transfer section is equal to one and the mole fractions of the vapour and liquid are identical in any cross section of the mass transfer section reducing the mass balance equation (3.32) to

$$y_i = x_i \tag{3.32a}$$

and the differential equation equation (4.14) to

$$\frac{dx_j}{dx_i} = \frac{(y_j^* - x_j)}{(y_i^* - x_i)} . \tag{4.14a}$$

4.2.1.1 Vector field of the differential equation

Equation (4.14a) belongs to the group of hypergeometric differential equations and has some interesting properties [25] as illustrated by the related vector field shown in Fig. 4.8 in the reference frame of a Gibbs triangle.

The vector field contains three singular solutions in the form of straight lines with two of them crossing each other alternately and thus forming a Gibbs triangle. Since both the nominator and the denominator of equation (4.14a) are zero at the intersections of the straight lines or corner points of the triangle, equation (4.14a) is not defined at these points and such points are called nodes or distillation endpoints with a different characteristic. Whereas in the neighbourhood of the left and the right node all vectors are pointing away or towards the node, respectively, the vectors at the top node are pointing in part towards and in part away from the node. Due to this behaviour the left node is called an unstable node, the right one a stable node and the top node a saddle point [25]. The straight lines defining the Gibbs triangle are so-called separation lines or separatrices [11–13] since the physically feasible solutions of equation (4.14a) are limited to the Gibbs triangle.

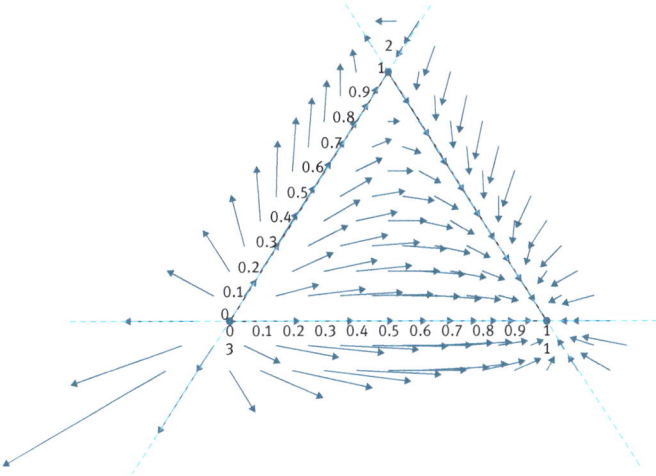

Fig. 4.8: Vector field of the distillation lines at total reflux (- - separation lines).

4.2.1.2 Field of distillation lines

For the total reflux case follows from equation (3.22)

$$\frac{dx_i}{y_i^* - x_i} = \frac{dx_j}{y_j^* - x_j} = d(NTU_V) \tag{4.15}$$

and from equation (3.25) and rearrangement of the two integrals one obtains the solution [12]

$$\frac{x_j}{x_i} = c_j \cdot \left(\frac{x_k}{x_i}\right)^{\frac{(\alpha_i - \alpha_j)}{(\alpha_i - \alpha_k)}} \tag{4.16}$$

with the constant c_j given by the initial conditions.

It should be noted that equation (4.16) in the light of equation (3.52) also applies to the simple distillation.

It is important to note that the characteristic variables of equation (4.16) are the mole ratios rather than the mole fractions which taking the light boiling component (1) as the reference component, the intermediate boiling component (2) as component j and the heavy boiling component (3) as component k yields

$$X_{21} = \left(\frac{x_2}{x_1}\right), \tag{4.17}$$

$$X_{31} = \left(\frac{x_3}{x_1}\right) \tag{4.18}$$

and

$$X_{21} = c_{21} \cdot X_{31}^{e_{21}} \tag{4.19}$$

with

$$e_{21} = \frac{(\alpha_1 - \alpha_2)}{(\alpha_1 - \alpha_3)} \, . \tag{4.20}$$

The constant c_{21} is obtained from the initial conditions.

The concentration profile $x_1 = f(x_2)$ is determined using the mass balance equation in the form

$$x_1 = \frac{1}{1 + X_{21} + X_{31}} \, . \tag{4.21}$$

Inserting equations (4.19) and (4.18) yields

$$x_1 = \frac{1}{1 + c_{21} \cdot X_{31}^{e_{21}} + X_{31}} \tag{4.22}$$

and

$$x_2 = X_{21} \cdot x_1; \quad x_3 = X_{31} \cdot x_1 \, . \tag{4.23}$$

The positive direction of a distillation line is defined in accordance with the positive direction of the driving force vector $[y^* - x]$, i.e. the distillation line starts at the bottom of a mass transfer section and moves upward to the top.

A graphical solution of equation (4.19) is derived using the Gibbs triangle with the right corner corresponding to the pure low boiling component (1), the top corner to the intermediate boiling component (2) and the left corner to the high boiling component (3) as shown in Fig. 4.9. The variable X_{31} is taken as the abscissa and the variable X_{21} as the ordinate. Since the variable X_{31} represents a straight line passing from the abscissa through the upper corner and the variable X_{21} a straight line starting from the ordinate and passing through the left corner of the Gibbs triangle, the intersection of these two straight lines gives the related mole fractions of the distillation line.

Starting from an initial concentration x_1, x_2 and drawing the two straight lines through this point and the left and upper corner of the Gibbs triangle, respectively, allows to calculate the constant in equation (4.19) and subsequently determine the variable X_{21} as a function of the variable X_{31}.

Varying the constant c_{21} from zero to infinity yields the total distillation lines. All distillation lines run between the pure components (1) and (3) with the sides of the Gibbs triangle corresponding to the limiting binary mixtures as shown in Fig. 4.9.

The relative volatilities used in this chapter relate to the ternary mixture methanol (1)–ethanol (2)–1-propanol (3) approximated as an ideal mixture with $\alpha_1 = 3.25$, $\alpha_2 = 1.90$ and $\alpha_3 = 1$.

The concentration profiles as a function of the dimensionless height and the number of theoretical stages calculated from equation (4.24) in combination with equation (4.19)

$$H_v^* = \ln c_{31} - \ln \left(\frac{1}{1 + c_{21} \cdot X_{31}^{e_{21}} + X_{31}} \right) + \frac{\alpha_1}{\alpha_3 - \alpha_1} \cdot \ln X_{31} \tag{4.24}$$

are shown in Fig. 4.10–4.12.

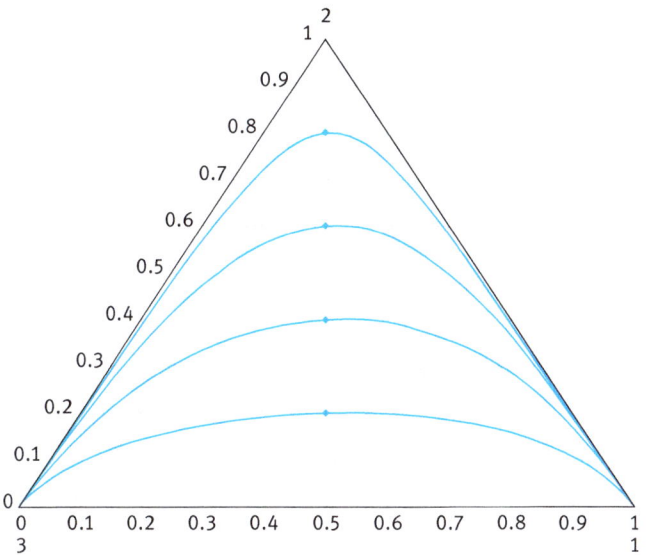

Fig. 4.9: Distillation lines at total reflux (○ initial conditions).

4.2.1.3 Effect of the mass transfer resistance

As discussed in Chapter 3, the mass transfer resistances affect the driving forces in the liquid and the vapour phase and are accounted for by a change in the relative volatilities. Whereas for the limiting case of a resistance exclusively in the vapour phase the conventional relative volatilities α_i are to be used in the equation (4.20) of the distillation line, in the limiting case of a resistance exclusively in the liquid phase the reciprocal values of the conventional relative volatilities $1/\alpha_i$ are to be applied. The theoretical stage concept corresponds to an equal distribution of the resistances in the two phases at total reflux conditions (see Fig. 3.3). Writing equation (3.29) additionally for the component j and by division of the two equations follows that equation (4.16) is valid also for the theoretical stage model if the relative volatilities in the exponent are replaced by their logarithms [35].

The effect of the different mass transfer models on the distillation lines is shown in Fig. 4.10 and 4.11 for the concentration profiles as a function of the dimensionless column height and on the number of theoretical stages in Fig. 4.12 [37].

Since the theoretical stage concept corresponds to a mass transfer resistance ratio of one, the distillation lines are bound by the limiting cases of a total mass transfer in the liquid and the vapour phase as shown in Fig. 4.13.

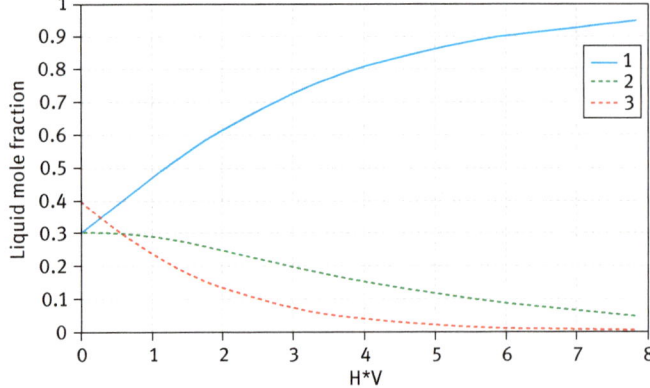

Fig. 4.10: Concentration profiles vs. the dimensionless height for a vapour phase located mass transfer resistance.

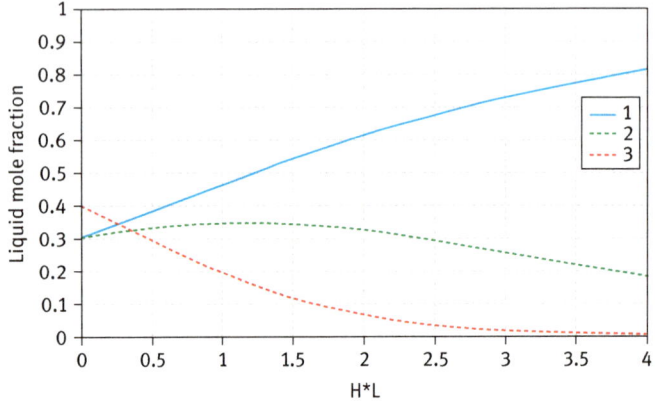

Fig. 4.11: Concentration profiles vs. the dimensionless height for a liquid phase located mass transfer resistance.

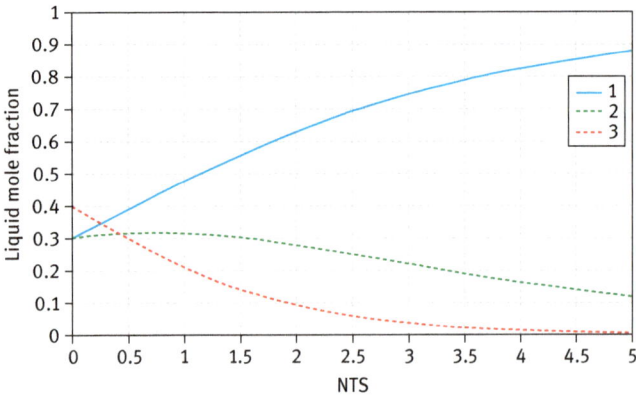

Fig. 4.12: Concentration profiles vs. the number of theoretical stages.

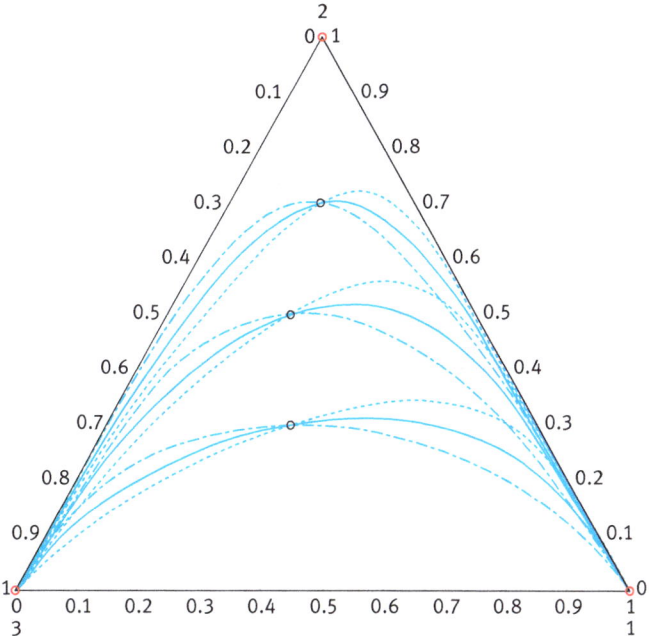

Fig. 4.13: Effect of the mass transfer resistance ratio on the course of the distillation lines (mass transfer resistance: - - - liquid, — theoretical stage, · · · vapour, ○ initial conditions).

For any distribution of the mass transfer coefficients

$$k = \frac{k_L}{k_V}$$

the exponent e of equation (4.19) may be approximated as

$$e = a + b \cdot \frac{1}{1 + k} + c \cdot \left(\frac{1}{1 + k}\right)^2 \tag{4.25}$$

with

$$a = e_G, \quad b = 4 \cdot e_{TS} - e_L - 3 \cdot e_G, \quad c = 2 \cdot e_L - 4 \cdot e_{TS} + 2 \cdot e_G \tag{4.26}$$

as shown in Fig. 4.14.

4.2.2 Distillation at partial reflux

As discussed above for total reflux, also at partial reflux the solutions of equation (4.14) contain three singular solutions in form of straight lines with two of them crossing each other alternately forming a displaced triangle as shown in Fig. 4.15. As both the

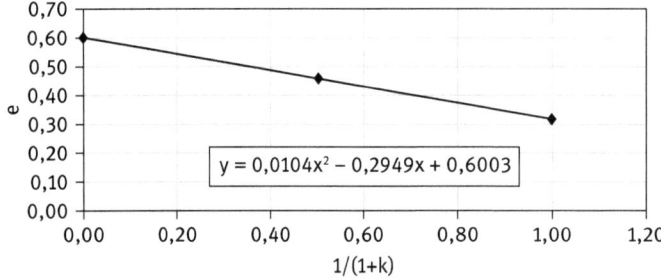

Fig. 4.14: Approximation of the power *e* vs. the ratio of the mass transfer coefficients (equation (4.25)).

nominator and the denominator of the differential equation (4.14) are zero at the intersections of the straight lines, the differential equation is not defined at these points and such points are called nodes. Since the feasible concentration profiles or distillation lines are limited to the triangle defined by the straight lines the triangle is named a distillation space and the straight lines are called separation lines. At total reflux the separation lines coincide with the sides of the Gibbs triangle and the distillation space becomes identical with the Gibbs triangle.

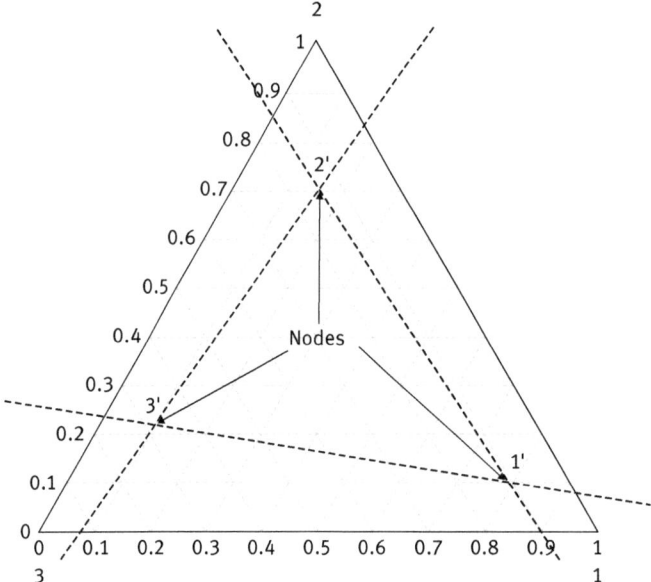

Fig. 4.15: Nodes and distillation space (- - - separation lines).

4.2.2.1 Vector fields at partial reflux

The distillation space of the concentration profiles of the liquid of a rectifying and a stripping section at partial reflux are visualized in the form of vector fields in Figs. 4.16 and 4.17, respectively, again indicating a different characteristic of the three nodes. All vectors $|y^* - y|$ close to the left node are pointing away from this node, the vectors near the top node are pointing partially towards and partially away from this node whereas the vectors near the right node are all pointing towards the node. Due to this behaviour the left node is called a stable node, the top node a saddle point and the right node an unstable node. Since in distillation the driving force vector $|y^* - y|$ is always directed towards a higher concentration of the lowest boiling component all distillation lines originate at the unstable node and end at the stable node.

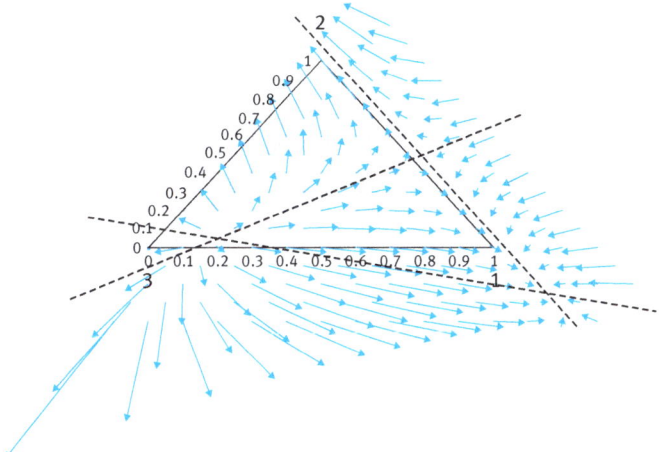

Fig. 4.16: Vector field of the liquid distillation lines of a rectifying section at partial reflux (- - - separation lines).

For each distillation space of the liquid at partial reflux there is also a distillation space of the vapour as shown e.g. in Fig. 4.18 depicting the vector field of the distillation lines of the vapour relating to Fig. 4.17.

The nodes of the distillation space of the vapour are in equilibrium with the nodes of the distillation space of the liquid.

4.2.2.2 Distillation spaces and separation lines

Like in binary distillation the nodes and the eigencoordinates of the distillation spaces follow from the condition $[y^* - y] = 0$, i.e. the driving forces of equation (4.14) vanish in the nodes. In order to obtain an analytical solution of the differential equation (4.14), the constants a_i and the flow ratio R in the mass balance equation (3.2) have to be

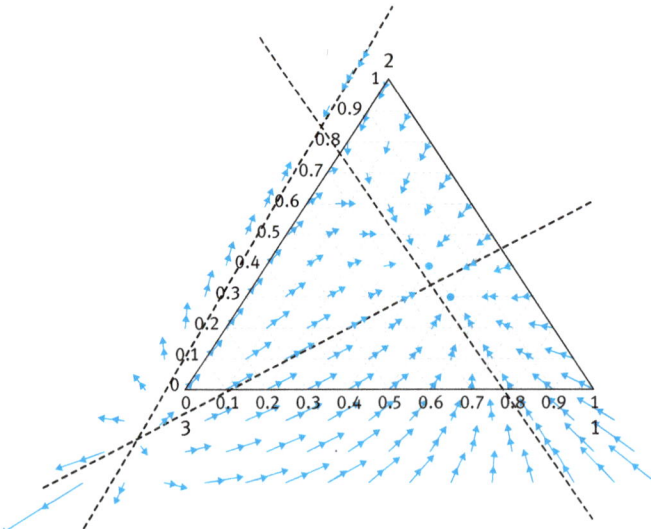

Fig. 4.17: Vector field of the liquid distillation lines of a stripping section at partial reflux (- - - separation lines).

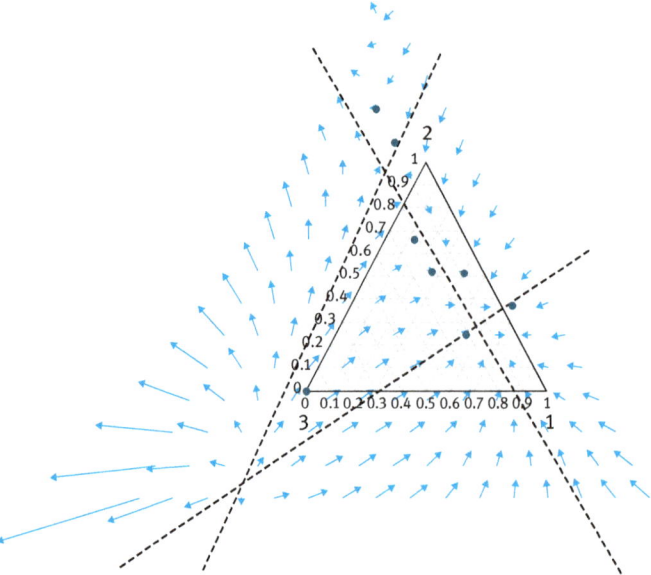

Fig. 4.18: Vector field of the vapour distillation lines of Fig. 4.17 (- - - separation lines).

removed. This is achieved by introducing a linear transformation of the coordinates x, y resulting in a new coordinate system ξ, η as discussed in Chapter 3 and shown in Fig. 4.19.

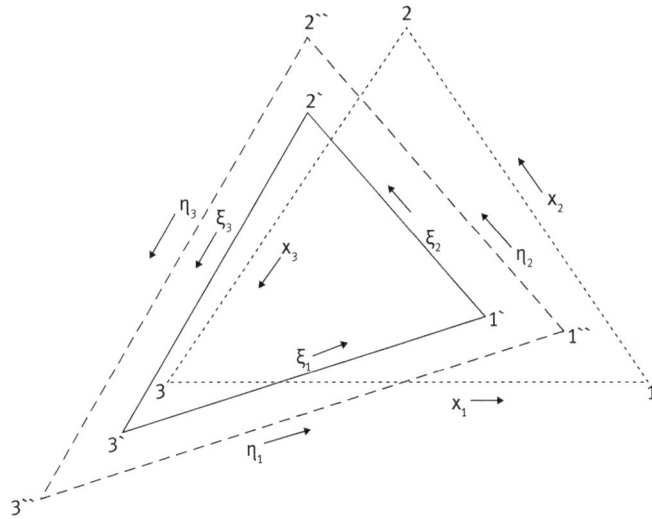

Fig. 4.19: Transformed distillation spaces (· · · Gibbs triangle, — distillation space of the liquid, - - - distillation space of the vapour).

The new ξ-coordinate-system of the distillation space of the liquid is defined by the separation lines representing the 'binary mixtures' of the distillation space with the coordinate ξ_1 being zero at the left node $3'$ and one at the right node $1'$, the coordinate ξ_2 being zero at the left node $3'$ and one at the top node $2'$ and the coordinate ξ_3 being zero at the top node $2'$ and one at the left node $3'$. The distillation space of the vapour with the η-coordinate-system is defined by the separation lines of the vapour having the property that all points of the vapour separation lines are equilibrium points of the separation lines of the liquid distillation space. It is noteworthy that the ratio of the length of one side of the distillation space of the vapour divided by the length of the corresponding side of the distillation space of the liquid is equal to the flow ratio L/V.

The composition of the nodes of the distillation spaces, the coordinates of the transformed spaces and the related composition profiles are calculated starting with the characteristic equation (3.43) as explained in Chapter 3.6.

As the solutions of the differential equation (4.14) are unique except in the nodes, it is impossible for any two distillation lines to cross each other except in a node. Since the straight line solutions of the differential equation (4.14), i.e. the separation lines are also distillation lines, they can not be crossed by another distillation line either except in the nodes.

From a distillation point of view, the nodes represent limiting points of the distillation lines, i.e. all distillation lines originate at the left (instable) node ($3'$), pass the saddle-point or node ($2'$) and terminate at the right (stable) node ($1'$). Thus, the triangle formed by the straight lines defines the distillation space and no distillation

is possible outside this space. The straight lines are called separation lines or separatrices, therefore.

As the driving force vector, i.e. the right side of the differential equation (4.14), is always directed towards a higher mole fraction of the low boiling component, this direction will be considered as the positive direction of the distillation lines.

It is important to note that there are three different kinds of separation lines:

Separation lines that pass from an unstable node to a stable node are called separation lines of the first kind. There is only one separation line of the first kind in a mixture with any number of components and the related nodes are the distillation endpoints.

Separation lines that pass from an unstable node to a saddle point or from a saddle point to a stable node are called separation lines of the second kind.

Separation lines that pass from one saddle point to another saddle point are called separation lines of the third kind. As they exist only for mixtures with four or more components, they indicate a distillation space of at least four components.

The separation lines are singular solutions of the differential equation (4.14) and obey the condition [13]

$$\frac{dy_j^*}{dy_i^*} = \frac{dx_j}{dx_i} = u \,.$$
(4.27)

Solving equation (4.27) by inserting the equilibrium equation (3.8) for mixtures with constant relative volatilities yields the quadratic equation

$$u = 0.5 \cdot (a \pm \sqrt{t^2 + 4 \cdot b})\,,$$
(4.28)

with

$$a = \frac{E_F \cdot (\alpha_i - \alpha_j) - \alpha_i \cdot x_{iF} \cdot (\alpha_i - \alpha_k) + \alpha_j \cdot x_{jF} \cdot (\alpha_j - \alpha_k)}{\alpha_i \cdot x_{iF} \cdot (\alpha_j - \alpha_k)}$$

and

$$b = \frac{\alpha_j \cdot x_{jF} \cdot (\alpha_i - \alpha_k)}{\alpha_i \cdot x_{iF} \cdot (\alpha_j - \alpha_k)}$$

indicating that there are always two separation lines passing through a ternary composition and that the slopes of the two separation lines are completely defined by the equilibrium. In addition to the liquid separation lines there are also separation lines of the vapour which run parallel to the separation lines of the liquid with each point of the vapour separation line being in equilibrium with its equilibrium point on the liquid separation line (see Fig. 4.20). It should be noted that there are two families of separation lines, i.e. one with a negative and one with a positive slope. Two separation lines of the family with a negative slope together with one separation line of the other family form a distillation triangle of the rectifying section and vice verse a distillation triangle of the stripping section. For a more extensive discussion of the nature of separation lines see [12, 32–34].

From a mathematical point of view the slopes of the separation lines passing through a node are related to the eigenvectors of the Jacobi-matrix of this node.

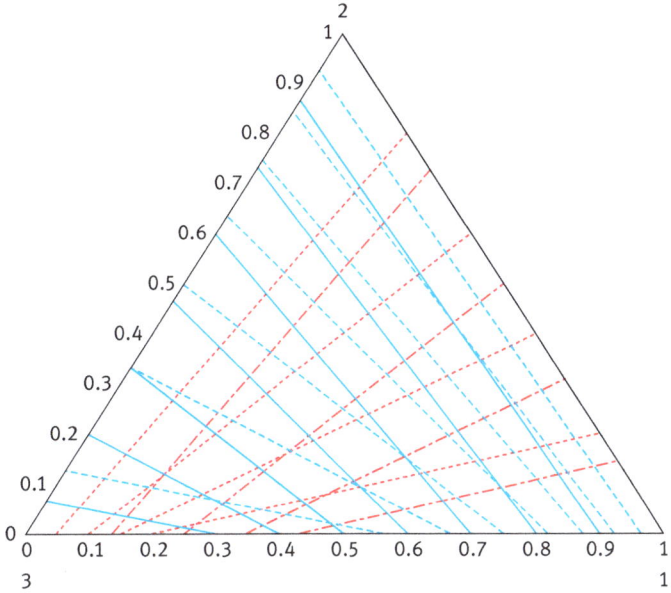

Fig. 4.20: Separation lines of an ideal ternary mixture (α = [3 2 1], —, - · · - liquid separation lines, - - -, · · · vapour separation lines).

It is important to note that the location of the nodes is uniquely defined by the intersections of the equilibrium equation (3.8) with the equation of the operating line (3.32) and does not depend on the kind of mass transfer model used.

4.2.2.3 Calculation of the distillation lines

The calculation of the distillation lines of the liquid follows the same procedure as outlined above for the distillation at total reflux, i.e.

– the α_i-values in equation (3.8) are replaced by the corresponding E_i-values,
– the mole ratios X are replaced by the mole ratios in the ξ-system defined as

$$\Xi_j = \frac{\xi_j}{\xi_i} \tag{4.29}$$

and

$$\Xi_k = \frac{\xi_k}{\xi_i} \tag{4.30}$$

so that the equation (4.19) reads

$$\Xi_j = \chi_j \cdot \Xi_k^{\lambda_j} \tag{4.31}$$

with

$$\lambda_j = \frac{(E_i - E_j)}{(E_i - E_k)} . \tag{4.32}$$

The constant in equation (4.31) is obtained by transforming the initial mole fractions from the x-space into the ξ-space by applying the inverse of the transformation equation

$$|x| = |r| \cdot |\xi|$$ (4.33)

with the elements of the transformation matrix $|r|$

$$r_{i,n} = \frac{a_i \cdot E_n}{\alpha_i - R \cdot E_n} .$$ (4.34)

Once the concentration profile is calculated in the ξ-space, equation (4.33) is used to transfer the fractions from the ξ-space into the x-space. The corresponding distillation lines of the vapour are then obtained by applying the balance equation

$$y_i = a_i + R \cdot x_i .$$ (3.2)

For calculating the concentration profiles of the liquid as a function of the dimensionless heights or the number of theoretical stages, the same equations are used as in the case of total reflux, however, converted to the ξ-space. E.g. in the ξ-space equation (3.25) reads

$$NTU_V = \ln c_{ki} - \ln \xi_i + \frac{E_i}{E_k - E_i} \cdot \ln \Xi_k = \frac{H_V}{(HTU)_v} = H_V^* .$$ (4.35)

The initial mole fractions in the ξ-space are obtained by transforming the mole fractions from the x-space using equation (3.36). After solving equation (4.35) the concentration profile has to be transformed into the x-space again using equation (4.33).

The same procedure applies to the case where the resistance is on the liquid side or to the theoretical stage concept by using the related equations as outlined above.

The equations developed in Chapter 3 are valid for systems with any number of components and can be directly applied to multicomponent systems. Since equation (4.14) represents an initial value problem, however, and as the required initial values can be determined from the feasible product spaces, as will be discussed later, it is more convenient to start with equation (3.43) in the form

$$V_R = \sum \frac{x_{D,i} \cdot D}{\alpha_i - R \cdot E_n} = \sum \frac{d_i}{\alpha_i - \Phi_{F,j}} ,$$ (3.43a)

where

$$d_i \le f_i = x_{F,i} \cdot F$$

is the recovery of the component i in the distillate and $\Phi_{F,j}$ the solution of equation (4.9) located between the alphas of the key components of the mixture to be separated.

Taking for example a ternary mixture with $\alpha = [3\ 2\ 1]$, a feed rate $F = 1$, a feed composition $x_F = [0.3\ 0.4\ 0.3]$, a thermal condition of the feed $q = 1$, a required recovery $d = [0.299\ 0.005\ 0.001]$ and a split (1)-2/2-(3), i.e the component 1 is the light

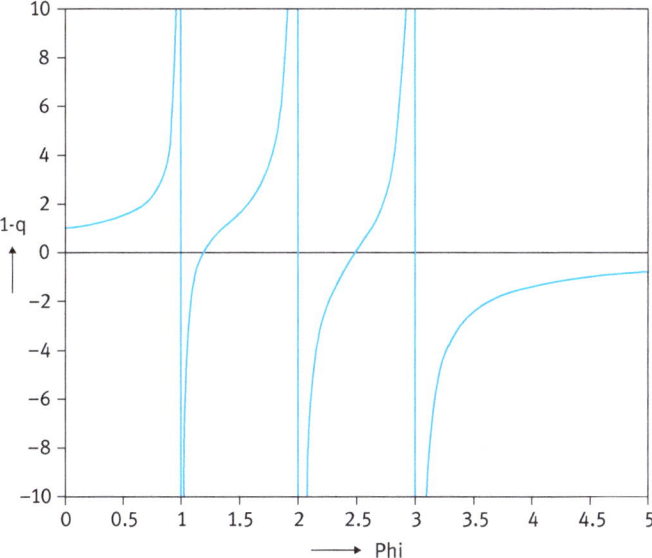

Fig. 4.21: Phi-function (equation (4.9), α = [3 2 1], x = [0.3 0.4 0.3], q = 1).

and the component 3 the heavy key component, then the roots Φ from equation (4.9) are Φ = [1.2 2.5] as shown in Fig. 4.21.

Taking a root with a value between the alphas of the key components, e.g. $\Phi_{F,2}$ = 2.5, the minimum vapour flow rate follows from equation (3.43a) to V_{min} = 0.833. With D = $\sum d = V_R - L_R$ the minimum flow ratio is R_{min} = 0.4. The maximum flow ratio of the stripping section is given by the mass balance (4.10c) to S_{max} = 1.6. The E-values of the nodes of the distillation sections follow from equation (4.8): E_R = [6.25 3.0 2.0], E_S = [2.00 1.56 0.75] with the E-value of the nodes $E_R(3)$ and $E_S(1)$ being equal to the E_F of the feed. The composition of the nodes follows from equations (3.40) and (3.41). These results are presented in Fig. 4.22 for the condition of limiting flows. The concentration profiles of Fig. 4.22 are given in Fig. 4.23 as function of the number of transfer units. The feed location corresponds to approximately NTU = 17.

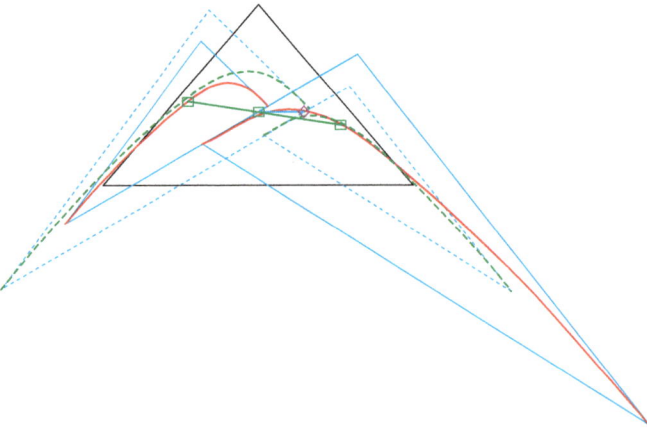

Fig. 4.22: Distillation spaces at limiting flow rates (— distillation spaces of the liquid, - - distillation spaces of the vapour).

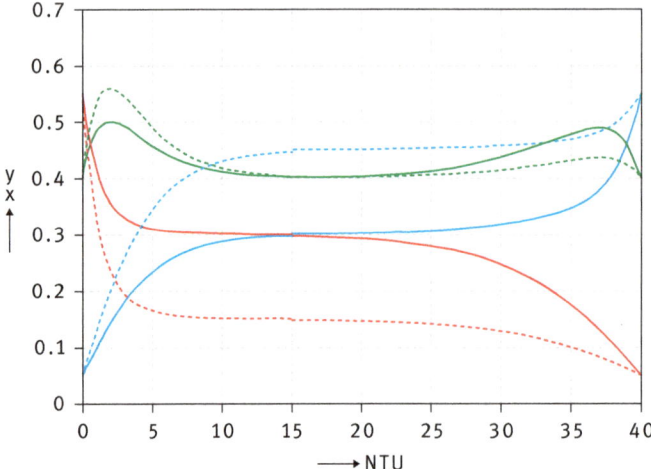

Fig. 4.23: Concentration profiles of Fig. 4.22 vs. the number of transfer units (— distillation lines of the liquid, - - distillation lines of the vapour).

4.2.3 Reversible distillation

Reversible distillation serves as a benchmark with respect to the energy consumption in distillation as it yields the minimum energy required in separating a mixture by distillation [28, 36]. The basic equation of reversible distillation, equation (3.56), was developed in Chapter 3

$$\alpha_i = E \cdot \left[(1 - R) \cdot \left(\frac{x_{i,0}}{x_i} - 1 \right) + 1 \right] . \tag{3.56}$$

Writing this equation also for component j and k and subtracting these from equation (3.56) results in

$$\alpha_i - \alpha_j = (1 - R) \cdot \left[\left(\frac{x_{i,0}}{x_i} - 1 \right) - \left(\frac{x_{j,0}}{x_j} - 1 \right) \right] \tag{3.56a}$$

$$\alpha_i - \alpha_k = (1 - R) \cdot \left[\left(\frac{x_{i,0}}{x_i} - 1 \right) - \left(\frac{x_{k,0}}{x_k} - 1 \right) \right]. \tag{3.56b}$$

Dividing these equations and rearranging yields

$$(\alpha_i - \alpha_j) \cdot \frac{x_{k,0}}{x_k} + (\alpha_j - \alpha_k) \cdot \frac{x_{i,0}}{x_i} + (\alpha_k - \alpha_i) \cdot \frac{x_{j,0}}{x_j} = 0 \tag{4.36}$$

and by multiplying with

$$\frac{1}{(\alpha_k - \alpha_j)} \cdot \frac{x_i}{x_{i,0}}$$

follows the more convenient form for calculating concentration profiles for the reversible distillation [14]

$$\frac{x_i}{x_j} = \frac{(\alpha_j - \alpha_k)}{(\alpha_i - \alpha_k)} \cdot \frac{x_{i,0}}{x_{j,0}} + \left(\frac{(\alpha_i - \alpha_j)}{(\alpha_i - \alpha_k)} \cdot \frac{x_{k,0}}{x_{j,0}} \right) \cdot \frac{x_i}{x_k}. \tag{4.37}$$

The distillation lines in Fig. 4.24 are obtained by applying equation (4.37) to the feed concentration and several concentrations on a product line coinciding with the equilibrium vector of the feed $[x_F \ -y_F^*]$ extending to the intersection of this line with the binaries 1–2 and 2–3, respectively. The distillation lines of the vapour are given by the concentrations in equilibrium with the concentrations of the liquid.

It is important to note that all distillation lines originating from a product line coinciding with an equilibrium vector pass through the feed composition or the equilibrium composition of the feed, respectively.

4.2.4 Product domains

4.2.4.1 Product domains in reversible distillation

As reversible distillation requires that no mixing takes place at the feed location i.e. the feed must have the same composition as the liquid and/or the vapour flows at the feed location, the distillation lines originating from feasible product compositions must pass through the feed composition. For a saturated liquid feed the feasible products meeting this restriction are limited to the equilibrium line, a line coinciding with the equilibrium vector of the feed $[x_F \ -y_F^*]$ and ranging from the feed composition to the intersections of this line with the binaries 1–2 and 2–3, respectively, as shown in Fig. 4.25. Since the liquid and the vapour at the feed location are in equilibrium,

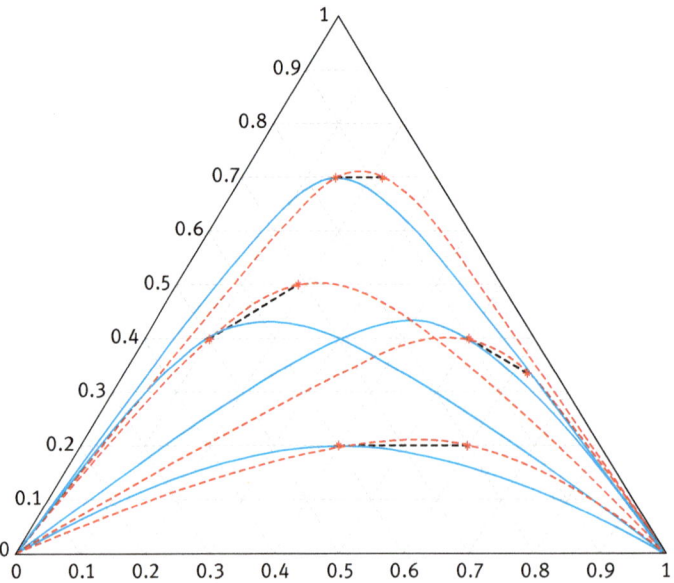

Fig. 4.24: Distillation lines at reversible distillation (full lines = liquid, dashed lines = vapour).

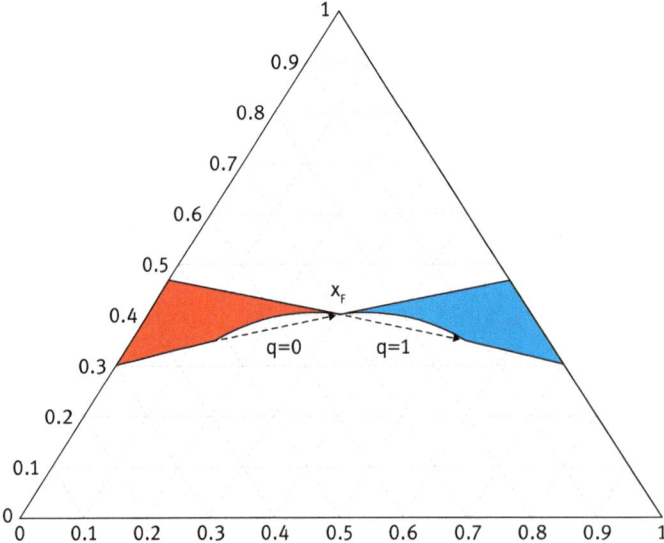

Fig. 4.25: Feasible product domains of the reversible distillation ($0 \leq q \leq 1$).

the feed is strictly limited to a thermal state $0 \leq q \leq 1$ and the feasible products of reversible distillation to the shaded areas of Fig. 4.25. These feasible splits restricted to thermal states $0 \leq q \leq 1$ are called reversible or transformation splits.

This restriction of feasible products is offset, however, if the condition of no-mixing at the feed location is relaxed and the dashed areas in Fig. 4.26 become available as feasible product domains. This is due to the property of the reversible distillation lines originating from the liquid feed and the composition of the vapour in equilibrium with the feed that any line coinciding with an equilibrium vector of these two lines $[y^* - x^*]$ passes through the feed composition x_F as indicated in Fig. 4.26.

The feasible product domains are given by the distillation line of the vapour, the two straight lines representing the mass balance restrictions valid for pure products and the sections of the side 1–2 and 2–3 of the Gibbs triangle representing products of a sharp separation. A sharp separation is defined as a split where at least one of the lighter boiling and at least one of the higher boiling components of the feed does not appear in the bottom product and distillate, respectively, like the splits (1)/(2)–3, 1–(2)/(3) or 1–(2)/(2)–3 in ternary distillation. The slash indicates the respective split and the components in parenthesis define the key-components.

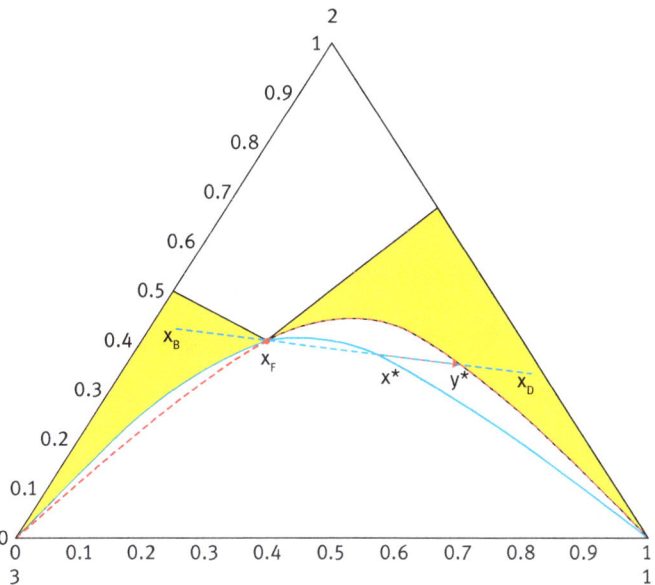

Fig. 4.26: Product domains of a relaxed reversible distillation (shaded areas = product domains).

The specific property of the reversible distillation that the reversible distillation lines originating from feasible product compositions must pass through the feed composition is a convenient tool to control whether assumed product compositions of an adiabatic distillation are feasible.

4.2.4.2 Product domains of the mass transfer and theoretical stage concept based distillation (adiabatic distillation)

The problem of solving the equations based on the mass transfer and theoretical stage concept imposes an initial value problem and requires to specify initial compositions which are most often the desired product compositions. Because of the limited degrees of freedom of a distillation column [17] predicting feasible product domains is of great importance, therefore. The feasible domains of an adiabatic distillation largely agree with the domains of the reversible distillation as discussed above except that the reversible distillation line through the feed composition is replaced by the distillation line at total reflux through the feed composition. Since the distillation lines at total reflux are affected by the distribution of the mass transfer resistances (see Fig. 4.13), the feasible product domains of the adiabatic distillation depend also on the distribution of the mass transfer resistances as shown in Fig. 4.27.

The feasible product domains are thus bound by the related distillation line through the feed composition at total reflux, the two straight lines representing the mass balance restrictions valid for pure products and the sections of the side 1–2 and 2–3 of the Gibbs triangle indicating the limiting products of sharp separations.

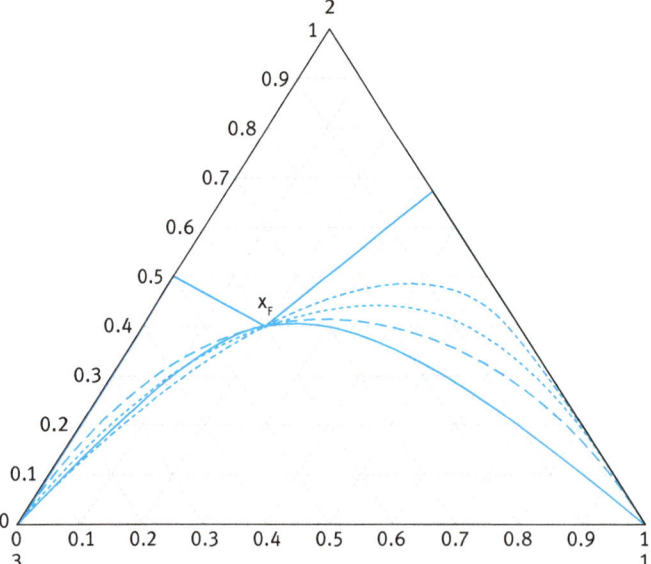

Fig. 4.27: Feasible product domains of the reversible and adiabatic distillation (— reversible line - - - total vapour resistance line ⋯ theoretical stage line - · - total liquid resistance line).

4.2.5 Distillation column

4.2.5.1 Distillation at the limiting flow ratios

The condition of the minimum flow ratio in the rectifying and the maximum flow ratio in the stripping section is characterised by an infinite number of transfer units or stages in the rectifying as well as the stripping section of a distillation column. The determination of these limiting flow ratios is explained on the basis of the possible splits of a ternary mixture into its products. In the limit there are three possible splits of a given ternary mixture as indicated by a slash and defined as

1. the transition split (1)–2/2–(3) where the product line coincides with the equilibrium vector of the feed producing a binary mixture 1–2 as the distillate and a binary mixture 2–3 as the bottom product,
2. the direct split (1)/(2)–3 where the light component 1 is obtained as the distillate and a binary mixture 2–3 as the bottom product and
3. the indirect split 1–(2)/(3) with the binary mixture 1–2 as the distillate and the component 3 as the bottom product.

In addition to these conventional definitions, a split is more exactly defined by the heavy and the light key component (shown above in parentheses) as will become apparent later in the context of multicomponent mixtures.

It should be understood that it is not possible to obtain a pure component by distillation as this would require a column of infinite height. Any product will thus contain impurities to a certain degree.

The limiting flow ratios for a transition split are given by

$$R_{\min} = \frac{(x_D - y^*)}{(x_D - x^*)} \tag{4.38}$$

and

$$S_{\max} = \frac{(x_B - y^*)}{(x_B - x^*)} \tag{4.39}$$

with $x^* = x_F$ and $y^* = y_F^*$ for $q = 1$.

For a ternary mixture with

$$\alpha = [3\ 2\ 1], \quad x_F = [0.2\ 0.3\ 0.5], \quad q = 1$$

the transition split is identical to the reversible split and the intersection of the product line with the binaries yields the limiting binary products

$$x_B = [0.000\ 0.231\ 0.769], \quad x_D = [0.571\ 0.429\ 0.000]$$

as shown in Fig. 4.28.

The limiting flow ratios from equations (4.38), (4.39) are $R_{\min} = 0.588$, $S_{\max} = 1.765$, respectively, and the corresponding distillation lines are given in Fig. 4.29, the

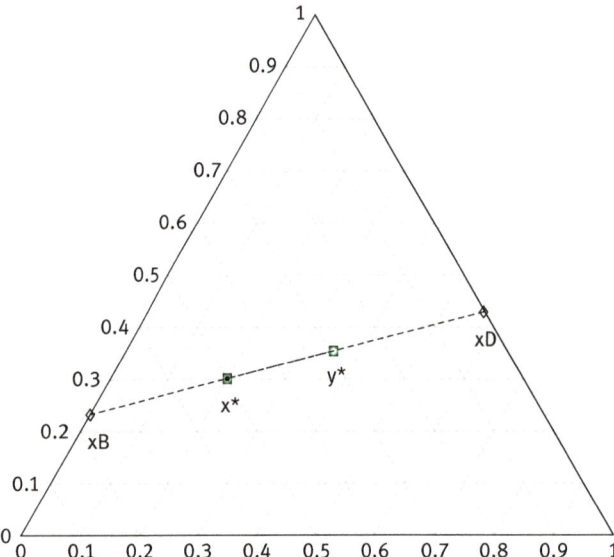

Fig. 4.28: Product line coinciding with the equilibrium vector $y^* = f(x^*)$ (for $q = 1$, $x_F = x^*$).

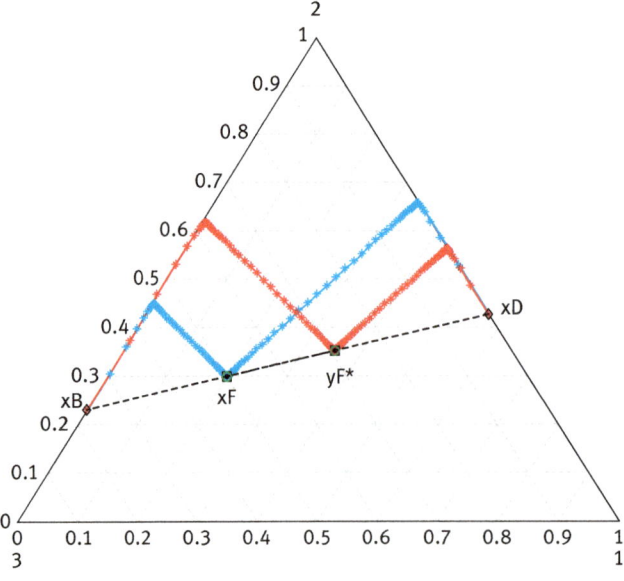

Fig. 4.29: Distillation lines at minimum flow ratio (transition split, full lines = liquid, dashed lines = vapour).

related distillation spaces in Fig. 4.30 and the concentration profiles as a function of the number of transfer units in Fig. 4.31.

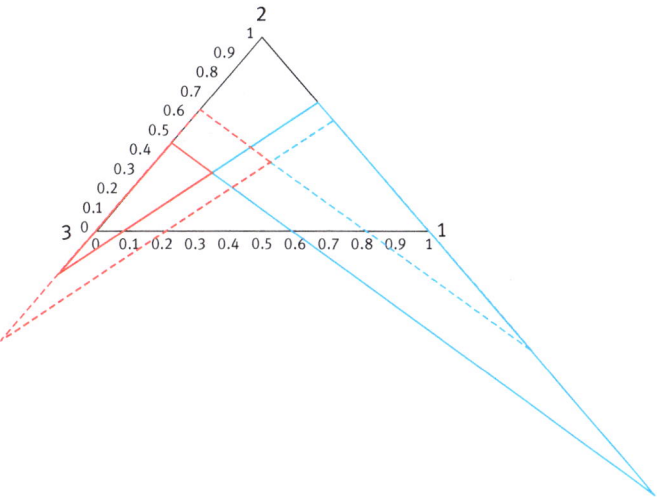

Fig. 4.30: Distillation spaces of Fig. 4.29 (x_F = [0.2 0.3 0.5])

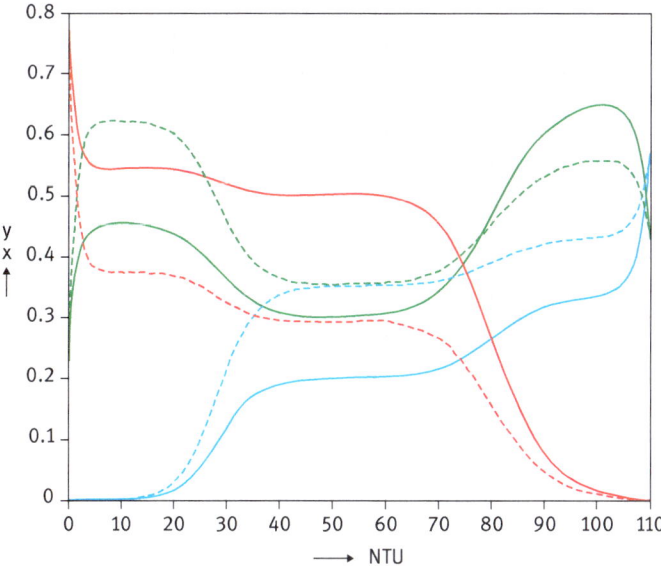

Fig. 4.31: Concentration profiles vs. the number of transfer units of Fig. 4.29
($R = R_{min}$, full lines = liquid, dashed lines = vapour)

It should be noted that the distillation spaces of the rectifying and the stripping section contact each other in Fig. 4.30 at the feed position (NTU = 50 in Fig. 4.31) and the
concentration in equilibrium with the feed composition exhibiting a zone of constant
composition or pinch zone as shown in Fig. 4.31, i.e. the two distillation spaces have

a common node coinciding with the feed composition. Since the products in Fig. 4.29 are specified as pure binaries, the concentration profiles must pass through a second zone of constant composition in each section of the distillation column and which are located in Fig. 4.31 at about $NTU = 10$ and $NTU = 100$, respectively. This fact allows to visualize the physical meaning of the roots Φ of equations (4.10a) and (4.10b). Taking a mass balance enclosing the equilibrium compositions of the pinch zone of the rectifying section and the composition of the distillate of Fig. 4.31 e.g. yields

$$V \cdot y^*_{\infty,i} - L \cdot x_{\infty,i} = x_{D,i} \cdot D \qquad (4.40)$$

which, taking into account equation (3.8), can be written as

$$V \cdot y^*_{\infty,i} = \frac{x_{D,i} \cdot D}{1 - L/V \cdot E_\infty/\alpha_{i,r}} = \frac{\alpha_{i,r} \cdot x_{D,i} \cdot D}{\alpha_{i,r} - R \cdot E_\infty} \qquad (4.41)$$

or by summation

$$V/D = \frac{1}{1 - R} = \sum \frac{\alpha_{i,r} \cdot x_{D,i}}{\alpha_{i,r} - R \cdot E_\infty} = \sum \frac{\alpha_{i,r} \cdot x_{D,i}}{\alpha_{i,r} - \Phi} \qquad (4.42)$$

indicating that Φ is the product of the flow ratio R times the averaged relative volatility E_∞ in the pinch zone of the rectifying section. A similar interpretation is valid for the stripping section.

For a transition split and $q \neq 1$ the equilibrium vector $[x^* -y^*]$ required in the equations (4.38) and (4.39) may be found either by calculating the reversible distillation lines of the liquid and the vapour originating from the chosen product compositions which intersect with the product line at the required equilibrium data (see Fig. 4.23) or by applying the equilibrium equation (3.8) along the chosen product line until the equilibrium data are located on the product line as shown in Fig. 4.32.

In practice it is impossible to implement splits yielding pure binaries as products since this would require an infinite height of both distillation sections as mentioned above. Therefore, all components will be present in the two products at least as an impurity without changing the basic behaviour of the split, however, as indicated e.g. by the transition split depicted in Fig. 4.33 and 4.34. For a transition split with $q = 1$ the limiting flow ratios can be directly calculated from the chosen feasible products and the calculated equilibrium compositions applying equations (4.38) and (4.39).

Since the equations (4.38), (4.39) apply to a transition split only, the limiting flow ratios of other splits must be calculated from the generally valid analytical method for ideal systems as developed by Underwood [10] based on the equation (4.9), (4.10a), and (4.10b) as derived in Chapter 4.1.

For a ternary mixture with $\alpha = [3\ 2\ 1]$, $x_F = [0.2\ 0.3\ 0.5]$ and $q = 1$ there are two roots of equation (4.9) i.e. $\Phi = [1.3758\ 2.5654]$. The calculation of the limiting flow ratios R_{min} and S_{max} by applying equations (4.10a) and (4.10b) depends on the kind of split with the concentrations of the distillate and the bottom product chosen from the feasible product domains (see Fig. 4.23).

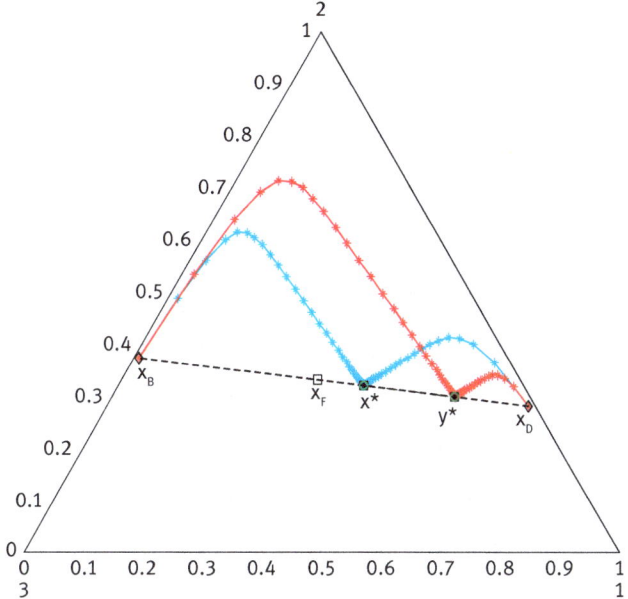

Fig. 4.32: Distillation lines of a transition split with an overheated vapour feed ($q = -0.5$).

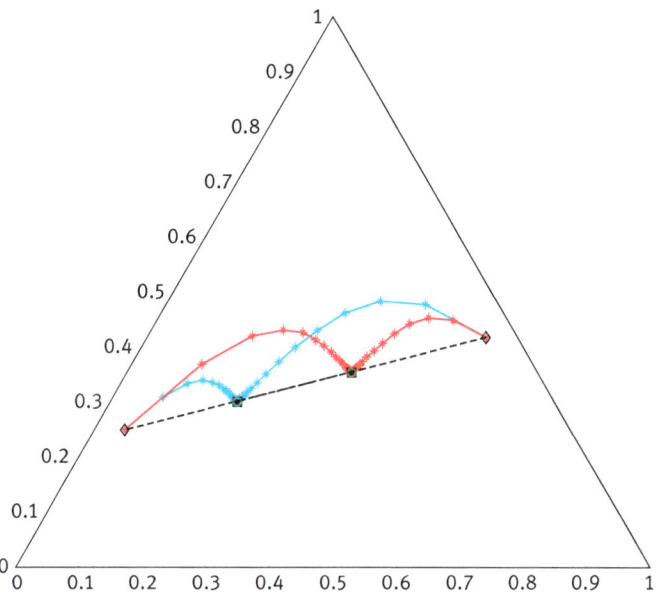

Fig. 4.33: Distillation lines at minimum flow ratio (transition or reversible split).

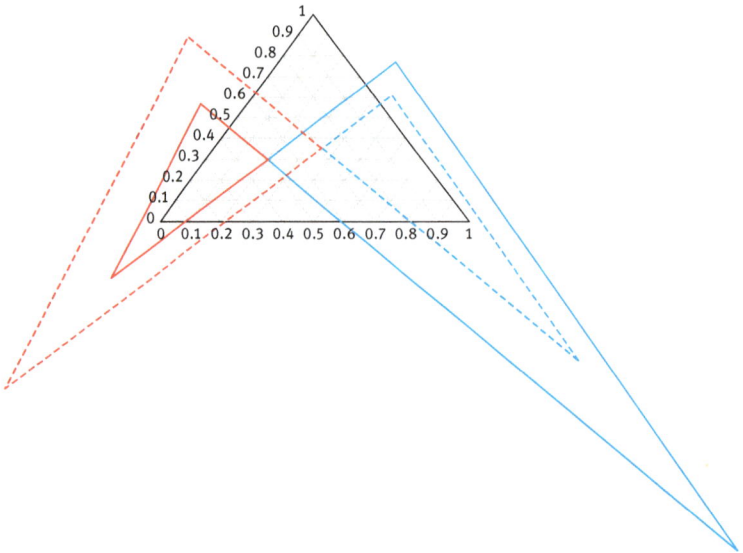

Fig. 4.34: Distillation spaces of Fig. 4.33.

In the case of the direct split (1)/(2)-3 the root with a value between the relative volatilities of the lowest (1) and the intermediate boiling component (2) is used in equation (4.10a). For a direct split with a chosen distillate composition of $x_{D,1} = 0.998$, $x_{D,2} = 0.001$ and a concentration of $x_{B,1} = 0.001$, $x_{B,2} = 0.3746$ of the bottom product and $\Phi = 2.5654$ as given above the minimum flow ratio follows from equation (4.10a) as $R_{min} = 0.853$ and the maximum flow ratio of the stripping section from equation (4.10b) as $S_{max} = 1.581$. The related distillation lines and spaces are given in Figs. 4.35 and 4.36.

For an indirect split 1–(2)/(3) the root with a value between the relative volatilities of the intermediate (2) and the highest boiling component (3) is used i.e $\Phi = 1.3758$. The figures corresponding to the following data

$$x_F = [0.20\ 0.30\ 0.50], \quad x_D = [0.90\ 0.05\ 0.05], \quad x_B = [0.05\ 0.35\ 0.60]$$

are given in Figs. 4.37 and 4.38.

It should be noted that for the direct and indirect split the distillation spaces have no common node at the feed composition but touch each other with one side of the distillation spaces by a common separation line.

In the case of the transition split (1)–2/2–(3) any of the two roots of equation (4.9) may be used in the equations (4.10a) and (4.10b).

The limiting flow ratios may also be determined by drawing the separation lines through the feed composition and the related equilibrium composition as given in Fig. 4.39. The limiting flow ratio R_d for a direct split is then given by the ratio of the distance of the distillate x_{Dd} from the separation line of the vapour with a positive slope

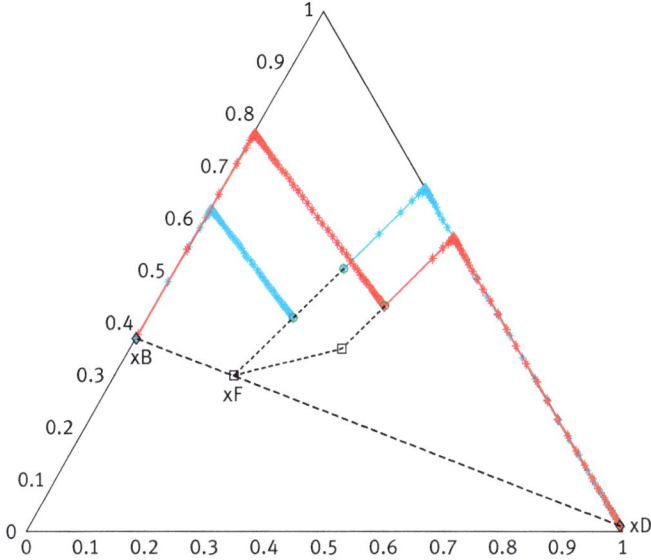

Fig. 4.35: Distillation lines at minimum flow ratio (direct split).

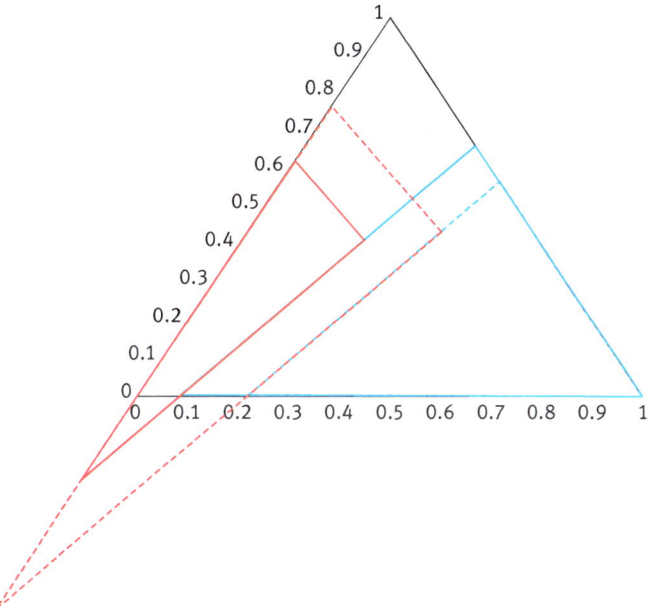

Fig. 4.36: Distillation spaces of Fig. 4.35.

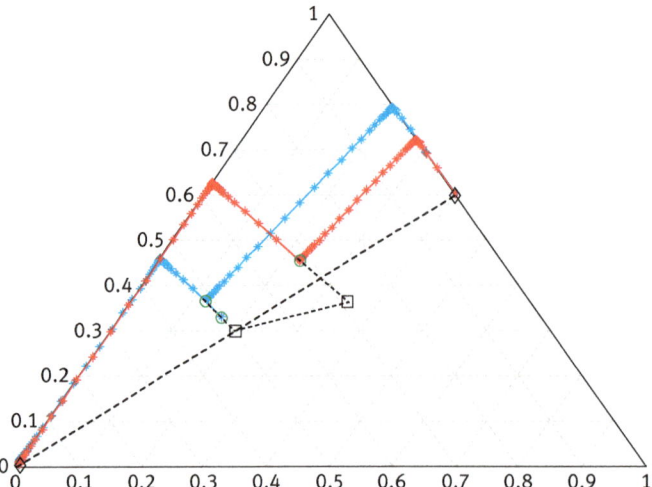

Fig. 4.37: Distillation lines at the limiting flow ratios (indirect split).

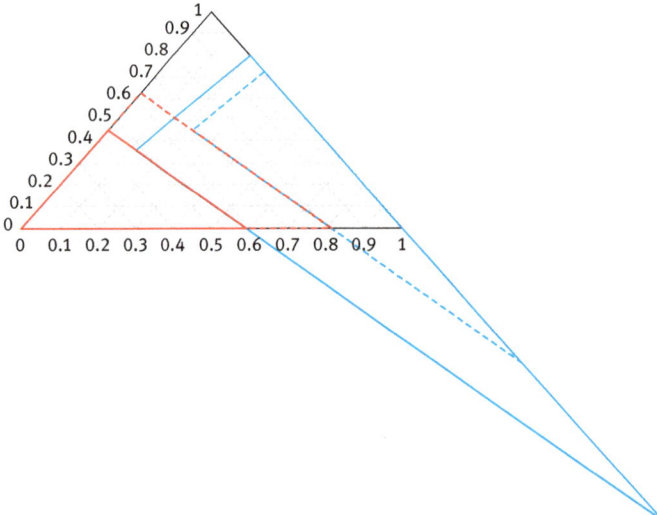

Fig. 4.38: Distillation spaces of Fig. 4.37.

divided by the distance of the distillate x_{Dd} from the separation line of the liquid with a positive slope. Similarly, the limiting flow ratio S_d for a direct split is given by the ratio of the distance of the bottom product x_{Bd} from the separation line of the vapour with a positive slope divided by the distance of the bottom product x_{Bd} from the separation line of the liquid with a positive slope. In the case of an indirect split, the limiting flow ratios are obtained by using the separation lines with a negative slope rather than the separation lines with a positive slope.

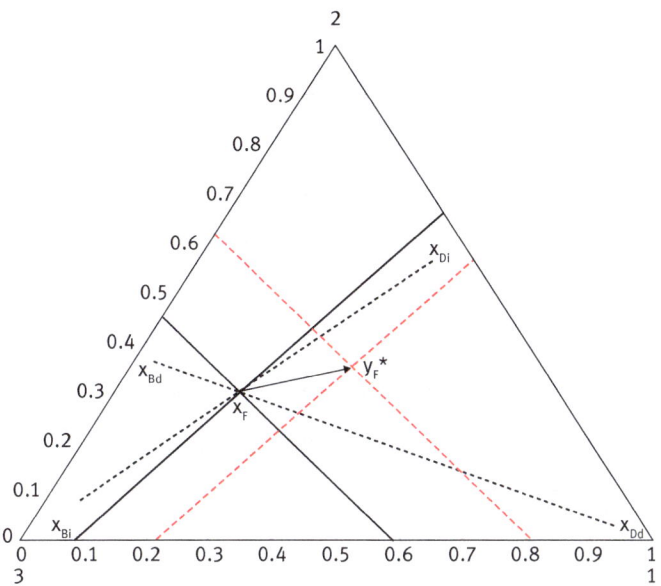

Fig. 4.39: Limiting flow ratios (— liquid separation lines, - - vapour separation lines, · · · product lines).

4.2.5.2 Distillation at non-limiting flow ratios

Once the product compositions have been chosen within the feasible product domains and the limiting flow ratios have been determined, the distillation lines can be calculated for the rectifying and the stripping section either by

1. using the theoretical stage concept or
2. by solving the differential equation (4.14) or
3. by applying the analytical solution of the differential equation as discussed in Chapter 4.2.2.

Since a distillation at the minimum flow ratio of the rectifying section and the related maximum flow ratio of the stripping section results in an infinite number of transfer units or theoretical stages, the optimal flow ratio, i.e $R > R_{\min}$ and $S < S_{\max}$ has to be established using criteria like ease of operation, flexibility concerning changing feeds or product compositions, energy consumption, integration into an existing distillation train a.s.o. [17].

The most common criterion is the "lowest total costs" obtained as the minimum of the sum of investment and operating costs. To this purpose the flow ratio of the rectifying section is increased which enlarges the energy requirements and thus the operating costs but reduces the number of transfer units or theoretical stages and thus the investment costs until the minimum total costs have been found.

4.2.5.3 Optimal feed location

A convenient procedure of how to determine the optimal feed location for given feed and product conditions and assumed flow ratios of the sections consists in first calculating the distillation lines originating from the product compositions up to their distillation endpoints and then using the mass balances over the feed section as discussed below.

Numerical procedures for calculating the optimum feed location of complex distillation columns are given in Adiche [37].

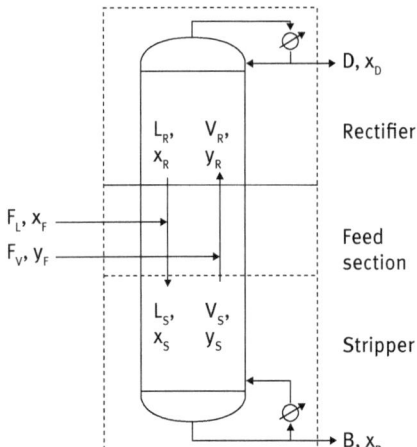

D, x_D

Rectifier

Feed section

Stripper

B, x_B

Fig. 4.40: Mass balances of the column.

Feed section:

$$F_L = L_S - L_R \tag{4.43}$$

$$F_V = V_R - V_S \tag{4.44}$$

$$F_L \cdot x_F = L_S \cdot x_S - L_R \cdot x_R \tag{4.45}$$

$$F_V \cdot y_F = V_R \cdot y_R - V_S \cdot y_S. \tag{4.46}$$

Rectifying section:

$$D \cdot x_D + V_R \cdot y_R = L_R \cdot x_R. \tag{4.47}$$

Stripping section:

$$B \cdot x_B + L_S \cdot x_S = V_S \cdot y_S. \tag{4.48}$$

1. The feed is a saturated liquid ($q = 1$)

   For a saturated liquid feed the vapour concentration profiles of the rectifier and the stripper intersect at the location $y_R = y_S$. Drawing a line from the composition of the distillate x_D through this intersection will intersect with the liquid concentration profile of the rectifier at the composition x_R as this line represents the mass balance of the rectifying section (4.48). Similarly drawing a line from the composition of the composition of the bottom product x_B through the intersection $y_R = y_S$ will intersect with the liquid distillation line of the stripper at the composition x_S as shown in Fig. 4.41.

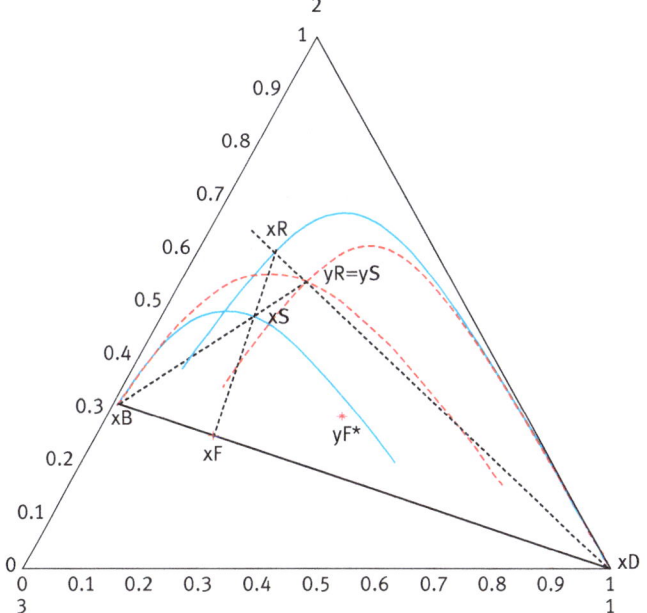

Fig. 4.41: Optimal feed location for a saturated liquid feed ($q = 1$).

2. The feed is a saturated vapour ($q = 0$)

   For a saturated vapour feed the liquid concentration profiles of the rectifier and the stripper intersect at the location $x_S = x_R$. Drawing a line from the composition of the bottom product x_B through this intersection will intersect with the vapour concentration profile of the rectifying section at the composition y_S as this line represents the mass balance of the stripping section (4.51).

   The line connecting the distillate x_D with the location $x_S = x_R$ intersects with the vapour distillation line of the rectifier yielding the concentration y_R at the feed location as shown in Fig. 4.42.

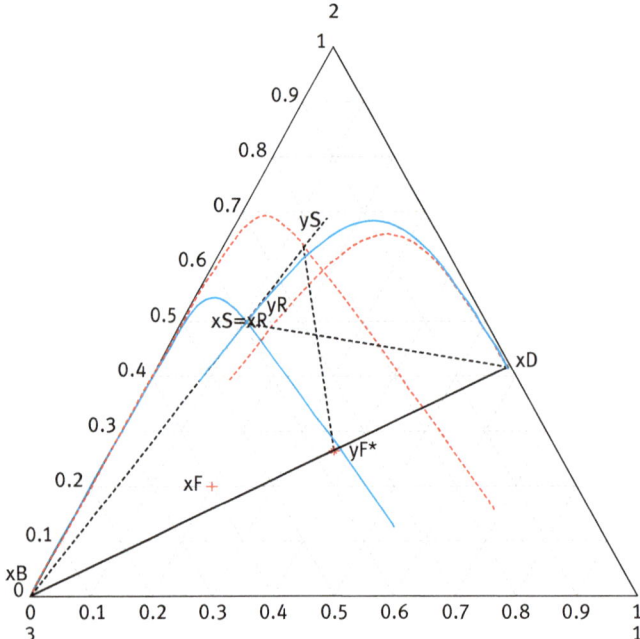

Fig. 4.42: Optimal feed location for a saturated vapour feed ($q = 0$).

3. The feed consists of two phases ($0 < q < 1$)
    For this case the optimum feed location is found by iteration only. An assumed
    mass balance line for the rectifying section (4.47) intersects with the vapour
    distillation line at the composition y_R and the liquid distillation line at the com-
    position x_R. Connecting this composition with the composition of the liquid
    feed x_F yields an intersection with the liquid distillation line of the stripper with
    the composition x_S corresponding to the mixing equation (4.45). Drawing a mass
    balance line from the bottom product through the composition x_S intersects with
    the vapour distillation line of the stripper at the composition y_S. The closing
    condition requires that the composition y_S must be located on the mixing line
    (4.46) connecting the composition y_R with the composition of the vapour phase
    of the feed y_F. If this condition is not fulfilled the procedure has to be repeated
    with another balance line of the rectifying section until the closing condition is
    met as shown in Fig. 4.43.

The analytical determination of the composition of the flows entering and leaving
the feed location follows from writing the equations (4.45) to (4.48) for all components and solving the resulting sets of equations for the unknown composition vectors
$[x_R \ y_R \ x_S \ y_S]$. This procedure applies also to distillation columns with multiple feeds
and side streams with any type of split [37].

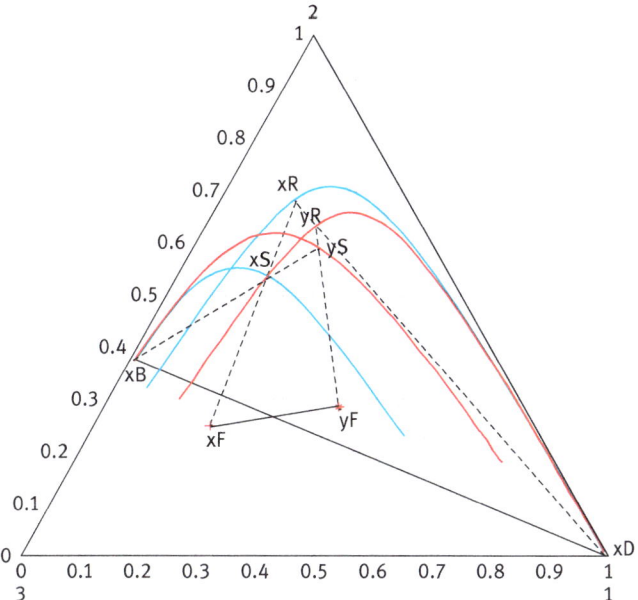

Fig. 4.43: Optimal feed location of a two-phase feed ($q = 0.5$).

It should be noted that the optimum feed location depends on the mass transfer model applied.

4.2.6 The fundamental equation of distillation

The important properties of reversible distillation:
1. Reversible distillation lines are completely determined by the thermodynamic properties of the mixture to be separated,
2. reversible distillation provides for the minimum possible energy for any distillation process,
3. all other distillation processes or distillation concepts are transforms of the reversible distillation equation,
4. the reversible distillation equation takes the simplest possible mathematical form,

suggest to term the equation of reversible distillation in the form [14]

$$(\varepsilon_i - \varepsilon_j) \cdot \Xi_k + (\varepsilon_j - \varepsilon_k) \cdot \Xi_i + (\varepsilon_k - \varepsilon_i) \cdot \Xi_j = 0 \qquad (4.49)$$

the "Fundamental Equation of Distillation" [14].

A comparison of equation (4.49) with equation (4.36) yields for the reversible distillation

$$\varepsilon = \alpha, \quad \varXi = \frac{x_0}{x}.$$ (4.50)

For the simple distillation and the distillation at total reflux follows from equation (4.20) for the limiting case of a resistance exclusively in the vapour phase

$$k = \infty, \quad \varepsilon = \alpha, \quad \varXi = \ln \frac{x_0}{x}$$ (4.51)

and the limiting case of a resistance exclusively in the liquid phase

$$k = 0, \quad \varepsilon = 1/\alpha, \quad \varXi = \ln \frac{x_0}{x},$$ (4.52)

whereas the distillation lines according to the theoretical stage concept with an even distribution of the mass transfer resistances are given by

$$k = 1, \quad \varepsilon = \ln \alpha, \quad \varXi = \ln \frac{x_0}{x}.$$ (4.53)

4.2.6.1 The line function of distillation

Since the equation (4.14) is an exact differential equation (see Appendix A.1) the above given solution (4.16) may also be interpreted on the basis of the vector field in Fig. 4.44 – in analogy to the "stream function" in fluid dynamics – as a "distillation line function". The distillation line function defines the distillation lines in the same way as the stream lines are the solutions of the stream function. The sides of the triangle are the separation lines of the distillation line function or the walls impermeable to the flow of the stream function whereas the instable and the stable node may be regarded as the source and the sink of the stream lines, respectively.

4.2.6.2 The potential function of distillation

The vector field of the potential function represents the orthogonal vector field of the distillation line function and is given in Fig. 4.44.

The analytical solution of the ternary potential function is

$$P = \left[(\alpha_i - \alpha_j) \cdot \left(\frac{x_j}{x_i} \right)^2 + (\alpha_i - \alpha_k) \cdot \left(\frac{x_k}{x_i} \right)^2 \right] \cdot 0.5.$$ (4.16a)

Potental lines corresponding to Fig. 4.44 calculated according to equation (4.16a) are given in Fig. 4.44a.

4.3 Quaternary mixtures

4.3.1 Distillation lines

The distillation lines of a quaternary mixture are calculated using two equations (4.16) with the mole ratio X_{14} as the independent and X_{24}, X_{34} as the dependent variables.

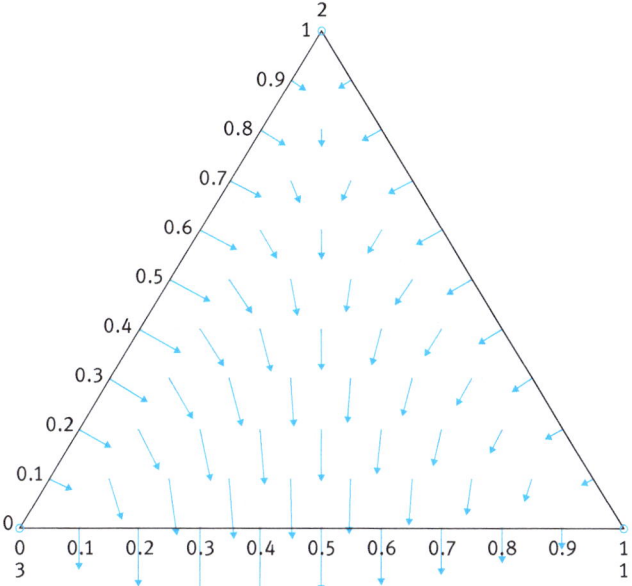

Fig. 4.44: Vector field of the potential function (α = [3 2 1]).

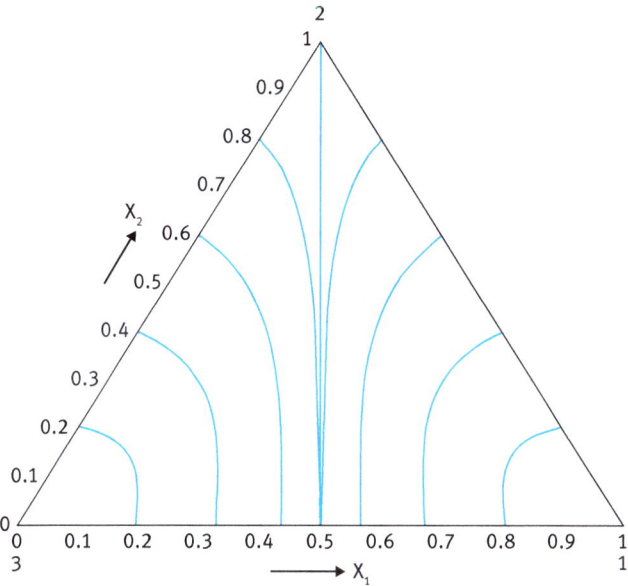

Fig. 4.44a: Potential lines of Fig. 4.44.

The distillation lines now take course within a tetrahedron as shown in Fig. 4.45 for total reflux and for partial reflux in Fig. 4.46.

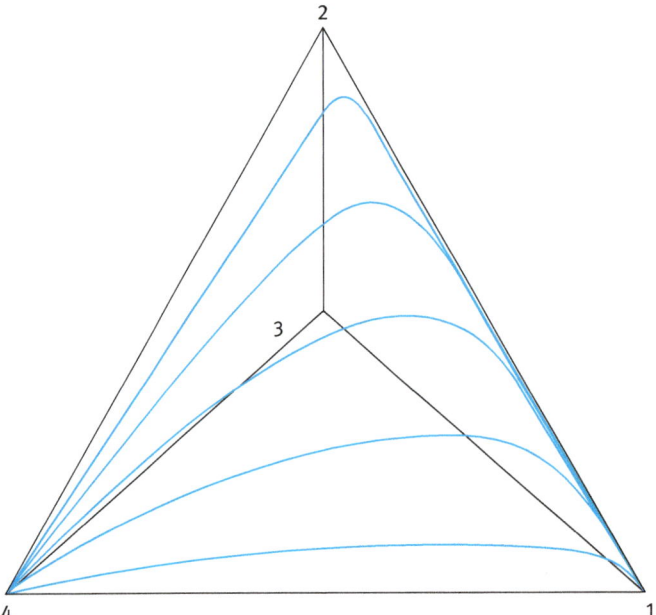

Fig. 4.45: Distillation lines of a quaternary ideal mixture at total reflux.

It is obvious that the distillation lines in Figs. 4.45 and 4.46 are not principally different from the distillation lines of a ternary mixture.

Since illustrating the distillation lines of multicomponent systems in form of a polyhedron reduces the information, a display of the concentration profiles as a function of the number of transfer units as given in Fig. 4.47 is used preferentially for systems with more than three components.

4.3.2 Product domains

4.3.2.1 Reversible distillation

The feasible product domains of a quaternary mixture for a given feed and a given thermal condition consist of the logical extension of the two one-dimensional feasible domains of a ternary mixture as depicted in Fig. 4.25 into two two-dimensional spaces as shown in Fig. 4.48 for a saturated liquid feed.

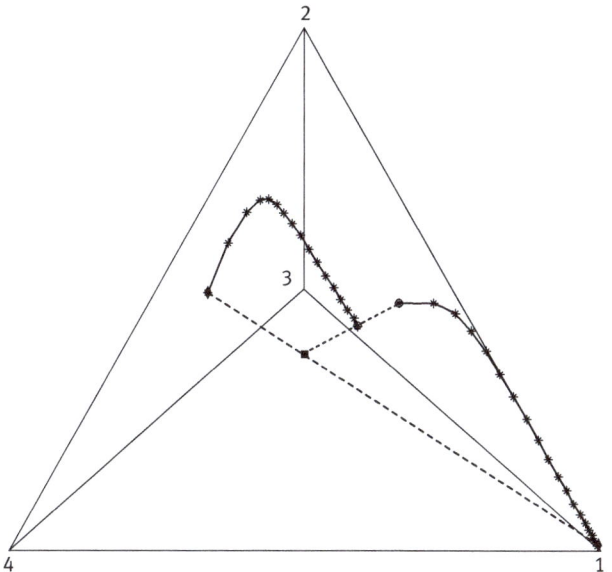

Fig. 4.46: Liquid composition profile of a mixture of methanol (1), ethanol (2), n-propanol (3) and i-butanol (4) (partial reflux, theoretical stage model).

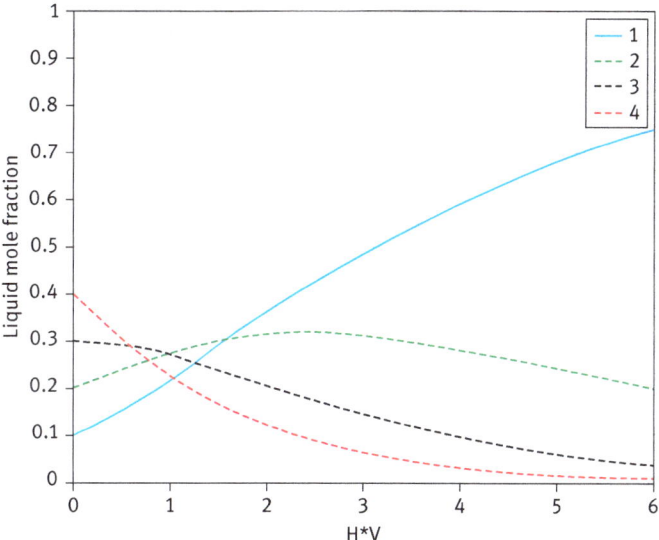

Fig. 4.47: Concentration profiles of a quaternary mixture vs. dimensionless height at total reflux (α = [4 3 2 1], $k = \infty$).

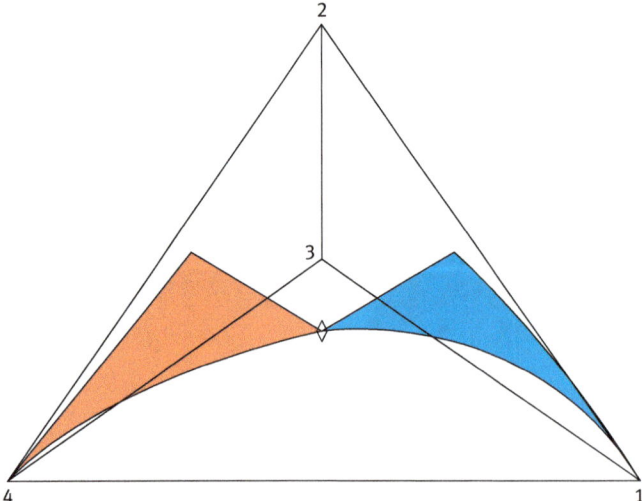

Fig. 4.48: Feasible product domains of the reversible distillation ($q = 1$).

4.3.2.2 Adiabatic distillation
4.3.2.2.1 Feasible product domains
The feasible product domains of an adiabatic distillation are given in Fig. 4.49 having the form of a distorted tetrahedron with the faces of the distillate domain defined by the spaces $x_F \rightarrow x_{123} \rightarrow DL \rightarrow 1 \rightarrow DL \rightarrow x_F$, $x_F \rightarrow x_{123} \rightarrow x_{12} \rightarrow x_F$, $x_F \rightarrow x_{12} \rightarrow 1 \rightarrow DL \rightarrow x_F$ and $x_{123} \rightarrow DL \rightarrow 1 \rightarrow x_{12} \rightarrow x_{123}$ and the bottom product accordingly by the spaces $x_F \rightarrow x_{234} \rightarrow DL \rightarrow 4 \rightarrow DL \rightarrow x_F$, $x_F \rightarrow x_{234} \rightarrow x_{34} \rightarrow x_F$, $x_F \rightarrow x_{34} \rightarrow 4 \rightarrow DL \rightarrow x_F$ and $x_{234} \rightarrow 4 \rightarrow DL \rightarrow x_{234}$. It should be noted that the lines $1 \rightarrow x_F$ and $x_F \rightarrow 4$ are three dimensional reversible distillation lines originating at the feed composition whereas the lines $x_{123} \rightarrow 1$ and $x_{234} \rightarrow 4$ are two dimensional reversible distillation lines belonging to the ternary mixtures x_{123} and x_{234}, respectively.

The points x_{123} and x_{234} are given by straight lines originating from the vertices 1 and 2, passing through the feed location and intersecting with the face 1–2–3 and 2–3–4 of the tetrahedron corresponding to the distillation cuts 1/234 and 4/123 of the feed x_F, respectively. Similarly, the points $x12$ and $x34$ are the products of the distillation cuts $3/x_{12}$ of the feed x_{123} and $2/x_{34}$ of the feed x_{234}, respectively.

4.3.2.2.2 Limiting flow ratios
The method of Underwood provides for a convenient method to determine the limiting flow ratios of ideal mixtures as explained in the following example. With a given saturated liquid feed ($q = 1$) and a feed composition $x_F = [0.25\ 0.25\ 0.25\ 0.25]$ follow from equation (4.9) the eigenvalues $\Phi = [3.4908\ 2.3278\ 1.1814]$. For feasible product compositions $x_D = [0.49\ 0.33\ 0.17\ 0.01]$ and $x_B(1) = 0.01$ and a transi-

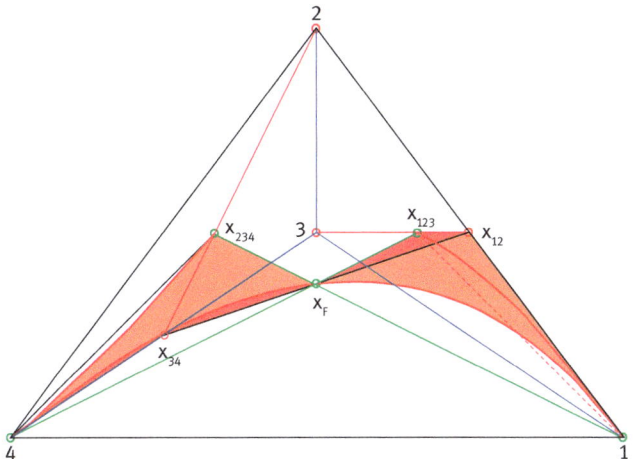

Fig. 4.49: Feasible product domains of a saturated liquid feed (x_F = [0.25 0.25 0.25 0.25], α = [4 3 2 1]).

tion split 1–(2)/(3)–4 equations (4.10a) and (4.10c) then give the minimum flow ratio in the rectifying section R_{min} = 0.375 and the maximum flow ratio in the stripping section S_{max} = 1.625, respectively. After determination of the nodes applying equation (3.40) the distillation spaces and the distillation lines can be calculated as explained in Chapter 3 and Chapter 4.4 with the results as shown in Fig. 4.50.

Since the solution as given in Fig. 4.50 is rather confusing and as the information given by the distillation spaces is not essential from a practical point of view, it is more convenient to numerically calculate the distillation lines and/or the concentration profiles (x, y) as a function of the number of transfer units (NTU) or theoretical stages (NTS).

For the above example and once the initial conditions have been established, the distillation lines are given by the set of two equations (4.14) in combination with the related mass balances (3.32) with the results shown in Fig. 4.51.

Similarly, the concentration profiles vs. the number of transfer units or theoretical stages follow from the set of two equations (3.22) and again in combination with the related mass balances (3.32) and taking into account the proper effect of the ratio of the mass transfer resistances. The concentration profiles vs. the number of transfer units are given in Fig. 4.52.

4.4 Multicomponent mixtures

The distillation of an ideal mixture with any number of components c is described by the same equations characterizing the distillation of an ideal ternary mixture as

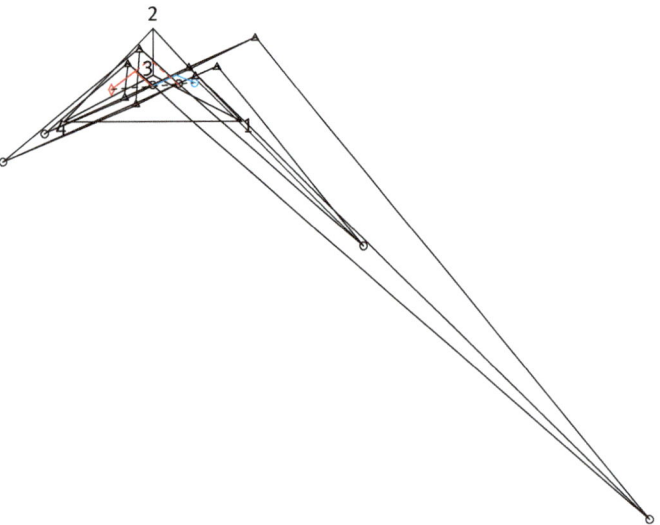

Fig. 4.50: Distillation lines and spaces of a quaternary mixture

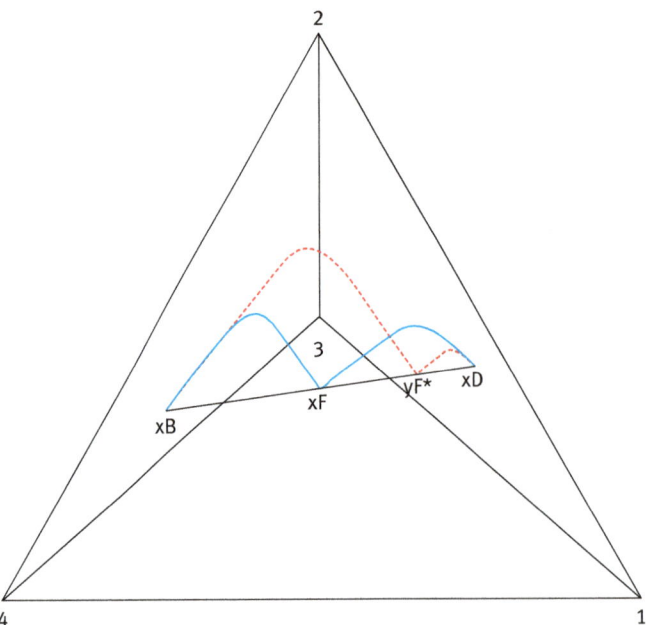

Fig. 4.51: Concentration profiles at the limiting flow ratios ($q = 1$, — liquid distillation lines,
$\cdots$ vapour distillation lines, - - - product line).

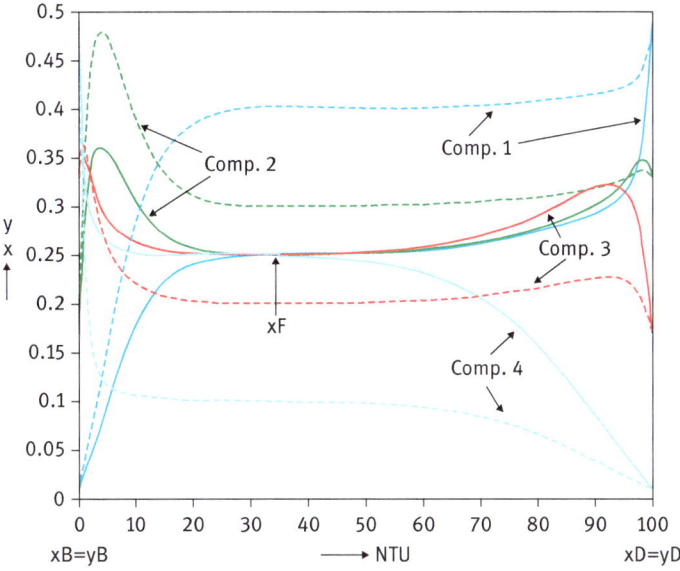

Fig. 4.52: Concentration profiles at the limiting flow ratios vs. the number of transfer units (q = 1, — liquid distillation lines, - - - vapour distillation lines).

all these equations are vector equations. For practical reasons it is recommended to order the components according to their increasing boiling point indexing the lowest boiling component as $i = 1$, the intermediate boiling components as $j = 2, 3, \ldots, c-1$ and the highest boiling component as $k = c$.

For a total reflux with $L/V = 1$ each equation of the set of $(c-1)$ equations (4.54) to (4.56) can be solved independently as a function of $x_k \cdot x_1^{-1}$ which strongly simplifies the calculation of the concentration profiles of a multicomponent mixture. It also indicates that a multicomponent mixture can be regarded as the superposition of the related ternary mixtures as disussed in [13, 14, 27]. The geometrical representation will become rather complex, however, and the course of the distillation lines will be illustrated best as a function of the number of transfer units or theoretical stages [17].

For a flow ratio $L/V \neq 1$ the distillation spaces of an ideal multicomponent mixture are obtained by a logical extension of the transformation procedure discussed above for a ternary ideal system and as explained in detail in [13].

4.4.1 Distillation at total reflux

For a mixture of k components there are $(k-2)$ intermediate boiling components and thus $(k-2)$ equations (4.16)

$$
\begin{aligned}
X_{21} &= C_{21} \cdot (X_{k1})^{e_{21}}, & e_{21} &= \frac{\alpha_1 - \alpha_2}{\alpha_1 - \alpha_k}, \\
X_{31} &= C_{31} \cdot (X_{k1})^{e_{31}}, & e_{31} &= \frac{\alpha_1 - \alpha_3}{\alpha_1 - \alpha_k}, \\
&\;\;\vdots & &\;\;\vdots \\
X_{k-1,1} &= C_{k-1,1} \cdot (X_{k1})^{e_{k-1,1}}, & e_{k-1,1} &= \frac{\alpha_1 - \alpha_{k-1}}{\alpha_1 - \alpha_k}.
\end{aligned}
\tag{4.54}
$$

It is important to note that the X_{ij} are invariant with respect to the number of components. Their value in a four component system e.g. is the same as in the related ternary subsystems. The constants C_{ij} are determined from the initial conditions and the set of equations (4.54) may be solved independently within the desired range of X_{k1} as a function of X_{k1}. The corresponding mole fractions x_i follow from the mass balance equation divided by x_1 and rearranged as

$$
x_1 = \frac{1}{\sum X_{ji}}
\tag{4.55}
$$

and with the other mole fractions calculated as

$$
x_i = X_{ji} \cdot x_1 .
\tag{4.56}
$$

The concentration profiles as a function of the number of transfer units for the case of a resistance only in the vapour phase is calculated using equation (4.57)

$$
NTU_V = \ln c_{ik} - \ln x_i + \frac{\alpha_{ik}}{\alpha_{ik} - 1} \cdot \ln X_{ik} .
\tag{4.57}
$$

For the case of a resistance in the liquid phase or the theoretical stage concept, the concentration profiles are calculated analogously (see Chapter 3).

The distillation space of a multicomponent mixture at total reflux is represented in principle by an irregular polyhedron with the frame of the polyhedron made up by the binaries of the mixture and the pure components as the vertices. The outer surface of the polyhedron consists of triangles representing the ternary mixtures of the total mixture. All possible distillation lines are contained within the polyhedron and extent from the highest boiling to the lowest boiling component.

As an example Fig. 4.53 shows the distillation space of a system with seven components in form of a distorted septahedron for better clarity. The septahedron consists of the dotted separation line of the first kind, the full separation lines of the second kind and the dashed separation lines of the third kind. The components 1, 2, 6 and 7 are located in one plane whereas the components 3, 4 and 5 project into the space

above the plane. The basic 3-component subsystems are 1–2–7, 1–3–7, 1–4–7, 1–5–7 and 1–6–7 each consisting of the lowest and highest boiling component of the mixture and one of the intermediate boiling components. The in reality six-dimensional distillation lines of Fig. 4.53 are given as projections on the 1–2–6–7-plane. The distillation lines originate at the point representing the highest boiling component 7 and end at the point representing the lowest boiling component 1.

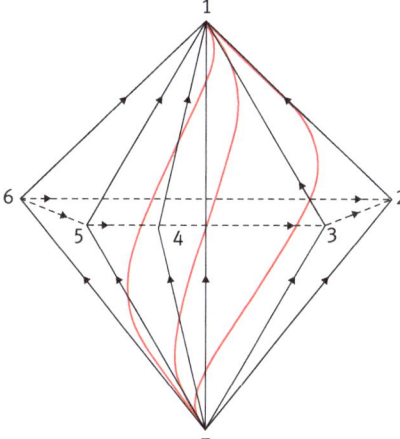

Fig. 4.53: Distillation space of a mixture with seven components.

As mentioned above, a more practical representation of higher-dimensional distillation lines is plotting the concentration profiles as a function of the number of transfer units or theoretical stages as shown in Fig. 4.54 for the above mixture with seven components at total reflux. The calculation follows the same procedures as outlined above in Chapter 3 using five equations of the form of equation (4.54) in combination with equation (4.57) for packed columns and equation (3.28) for columns with theoretical stages, respectively.

4.4.2 Distillation at partial reflux

At partial reflux the distillation takes place in two overlapping polyhedrons, one for the rectifier and one for the stripper and the design of a multicomponent distillation follows in principle the same steps as outlined in Chapter 3 and Chapter 4 for the distillation of a ternary or quaternary mixture, respectively. Because of the difficulty, however, to obtain product compositions from a highly dimensional feasible distillation space as required to establish the limiting flows from the Underwood equation, the iterative procedure as proposed by Shiras et al. [38] is a more convenient way to establish feasible product compositions.

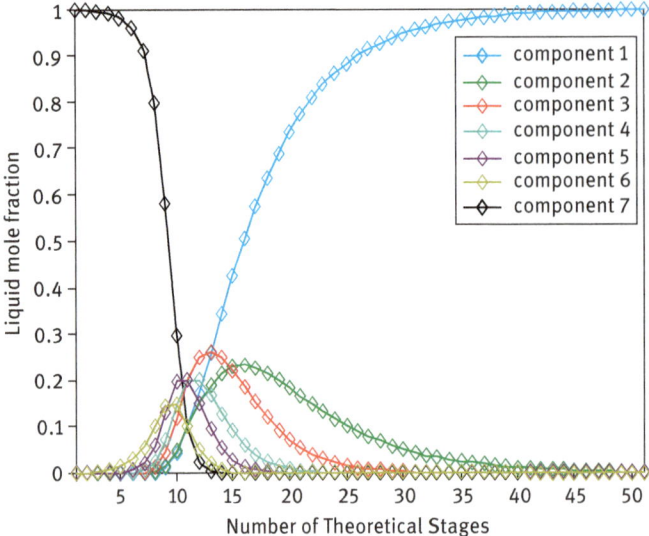

Fig. 4.54: Concentration profiles of a seven component mixture at total reflux vs. the number of theoretical stages.

The principle method of calculating the concentration profiles of a multicomponent distillation is presented in the Appendix A.2 taking a five component mixture as an example.

4.4.2.1 Product compositions
The procedure of Shiras et al. is based on the findings of Underwood [9, 10] that at limiting flow conditions the characteristic function of the rectifying section (4.4a) and the stripping section (4.4b) must have a common root Φ and that the numerical value of this root is located between the relative volatilities of the light and the heavy key-component defining the split of the mixture to be separated. From the condition of a common root for both distillation sections follows equation (4.9) so that for a given feed all roots Φ can be calculated. For a mixture with c components there are always $(c - 1)$ roots.

With setting the recovery of the two key-components in the distillate as

$$rec(i) = \frac{x_{Di} \cdot D}{x_{Fi} \cdot F} = \frac{d_i}{f_i} \tag{4.58}$$

equation (4.10a) takes the form

$$V_R = \sum \frac{\alpha_i \cdot d_i}{\alpha_i - \Phi}. \tag{4.10d}$$

Inserting the $(c - 1)$ roots Φ into equation (4.10d) and taking into account the two known values of d_i yields a set of $(c - 1)$ equations with $(c - 1)$ unknowns, i.e. V_R and $(c - 2)$ values of d_i.

If the solution of the set of $(c - 1)$ equations (4.10d) contains an unrealistic value of $d_i < 0$ or $d_i > f_i$ this d_i is set $d_i = \varepsilon$ or $d_i = f_i - \varepsilon$, respectively, and this equation is deleted from the set of equations. Then the set of the $(c - 2)$ remaining equations is solved and the above outlined procedure repeated until all unknowns have been obtained. ε should be set to a small but realistic value as $\varepsilon = 0$ could lead to improper initial values in calculating the concentration profiles. With the known limiting flow rate of the vapour V_R and the flow rate of the distillate D as the sum of all the flow rates of the components in the distillate d_i all other variables like the flow rates and the composition of the bottom product as well as the limiting flow ratios $R_{\min}$ and $S_{\max}$ can be calculated.

With the known initial conditions the concentration profiles can be calculated by using the theoretical stage concept or by solving the differential equations (4.15) numerically or by applying the theoretical concept based on the roots of the characteristic function as discussed in Chapter 4.2. The calculation of the optimum feed location and the determination of the optimal flow ratios applying appropriate optimisation criteria then finalizes the design of the distillation column.

4.4.2.2 Sequencing of distillation columns

Optimizing the separation sequence in distilling a polynary mixture is a rather complex problem and sophisticated methods have been developed to find the optimal sequence [18].

The minimum energy required for any distillation process is obtained if the process is operated at reversible conditions, i.e. the concentrations of the liquid and the vapour are in equilibrium in any cross section of the distillation column [28]. The feasible product compositions at reversible distillation are limited to a mass balance line coinciding with the equilibrium vector of the feed composition with the limits given by the intersection of this line with the boundaries of the distillation space as indicated in Fig. 4.25. I.e. only the heavy key component and all components with a relative volatility larger than the volatility of the heavy key component will appear in the distillate and only the light key component and all components with a relative volatility lower than the volatility of the light key component will appear in the bottom product. The composition of the products can thus be calculated from the intersections of the mass balance line with the limit of the distillation space.

Figure 4.55 gives an example for the energy optimal distillation sequence of a four component mixture with the limiting flow ratios given in Table 4.1 and the various distillate and bottom product compositions listed in Table 4.2.

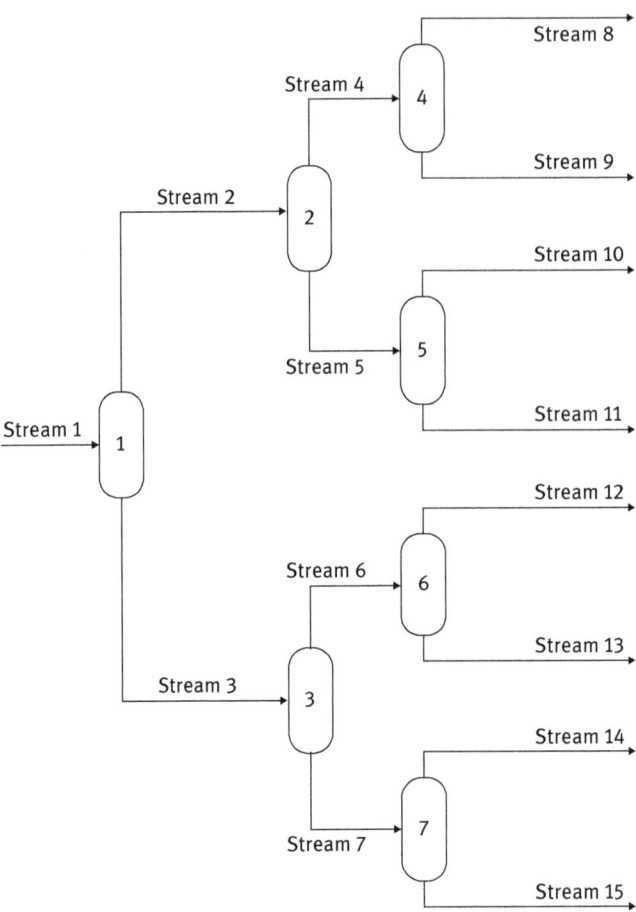

Fig. 4.55: Distillation sequence separating a four component mixture (Data see Table 4.2).

Tab. 4.1: Limiting flow ratios of Fig. 4.55.

Column	Minimum reflux ratio	Maximum reboil ratio
1	0.40	1.60
2	0.65	1.27
3	0.60	1.80
4	0.83	1.10
5	0.65	1.27
6	0.83	1.25
7	0.80	1.60

Tab. 4.2: Stream data of Fig. 4.55.

Stream	Flow rate	Quality	X_1	X_2	X_3	X_4
1	1	Liquid	0.250	0.250	0.250	0.250
2	0.50	Vapour	0.500	0.333	0.167	0
3	0.50	Liquid	0	0.167	0.333	0.500
4	0.356	Vapour	0.700	0.300	0	0
5	0.144	Liquid	0	0.419	0.581	0
6	0.166	Vapour	0	0.500	0.500	0
7	0.333	Liquid	0	0	0.250	0.750
8	0.250	Vapour	1	0	0	0
9	0.106	Liquid	0	1	0	0
10	0.061	Vapour	0	1	0	0
11	0.083	Liquid	0	0	1	0
12	0.083	Vapour	0	1	0	0
13	0.083	Liquid	0	0	1	0
14	0.083	Vapour	0	0	1	0
15	0.250	Liquid	0	0	0	1

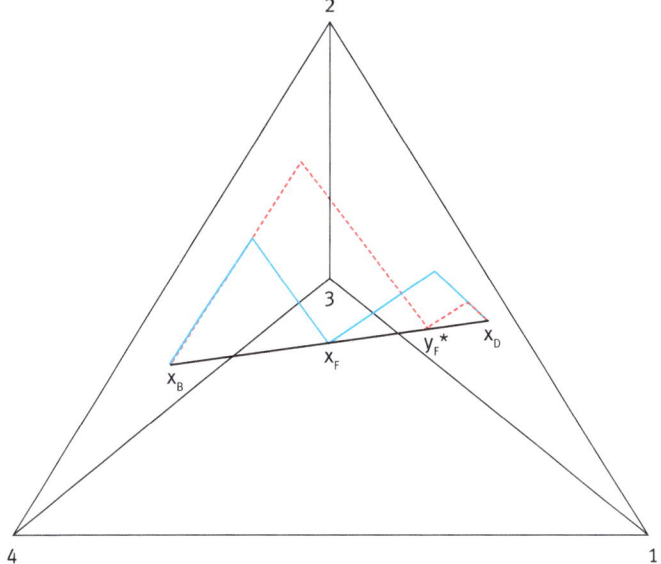

Fig. 4.56: Distillation lines of column 1 (α = [4 3 2 1]).

In column 1 the quaternary mixture is separated in two ternary mixtures as shown in Fig. 4.56. In the next columns 2 and 3 these ternary mixtures are reduced to binary mixtures with the distillation lines given in Figs. 4.57 and 4.58, respectively. The columns 4 to 7 finally produce the pure components 1 to 4.

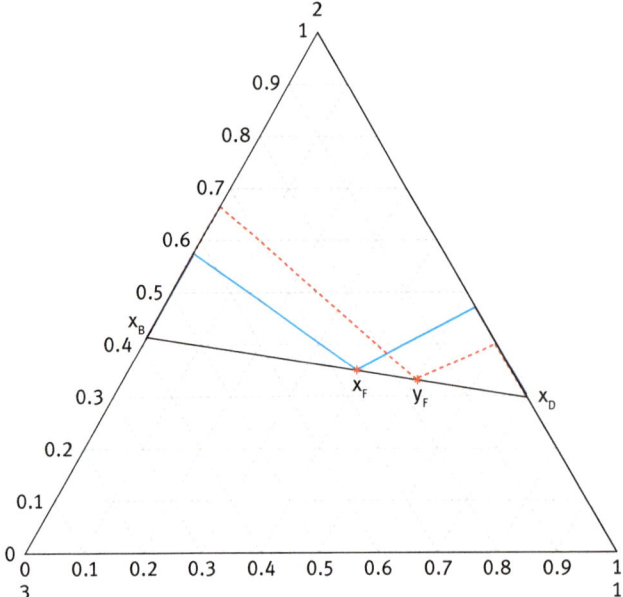

Fig. 4.57: Distillation lines of column 2 (α = [4 3 2]).

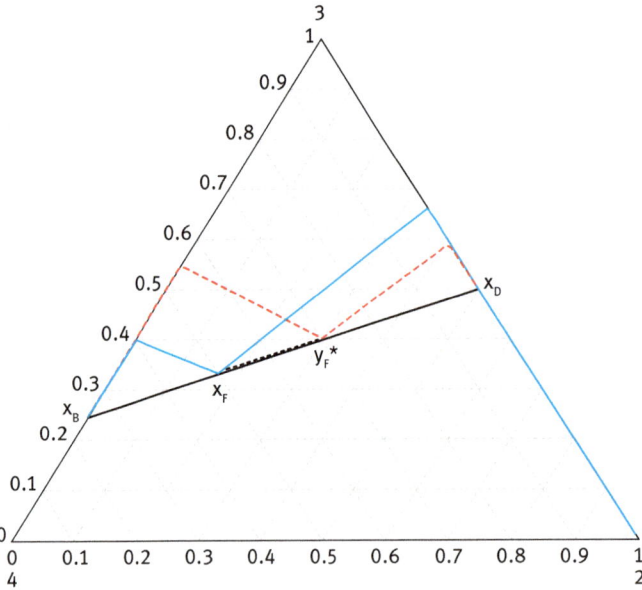

Fig. 4.58: Distillation lines of column 3 (α = [3 2 1]).

As the reversible distillation process requires columns of infinite height it can not be implemented technically but serves as a reference process only with respect to the minimal energy required in a distillation process.

In practice, there are different feasible distillation sequences like separating successively the lowest or the highest boiling component of a mixture which will require $(nc-1)$ columns for a mixture with nc components. Another possibility is to split a four component mixture e.g. into a distillate containing the two lowest boiling components and a bottom product with the two highest boiling components and subsequently separating the two binary mixtures into pure components which again will require $(nc-1)$ columns.

The number of such feasible separation sequences is given as

$$Ss = \frac{[2 \cdot (nc-1)]!}{(nc-1)! \cdot nc!} . \tag{4.59}$$

Whereas there are five feasible separation sequences for a four component mixture, the separation sequences for an eight component mixture increase already to 429 feasible separation sequences.

In order to handle this exponentially increasing separation sequences heuristic rules have been developed. Heaven [56] e.g. identified four general sequence selection heuristics based on energy consumption considerations:

1. Separations where the relative volatility of the key-components is close to unity should be performed in the absence of non-key components.
2. Sequences which remove the components one by one in column overheads should be favored.
3. Separations which give a more nearly equimolal division of the feed between the distillate and bottoms product should be favored.
4. Separations involving very high specified recovery fractions should be reserved until last in a sequence.

Since the above heuristics may conflict with one another, several sequences should be examined in a specific case in order to find the dominant of these heuristics [37] or more sophisticated methods should be applied [18].

4.5 The conceptual design of the distillation of multicomponent mixtures

4.5.1 Initial conditions

1. Completely define the feed.
2. Collect the parameters required for the VLE-equilibria calculations.
3. Define the split and the light and heavy key-component.

Define the recovery of the light and heavy key-component in the distillate and the bottom product, respectively.

4. Define the distribution of the mass transfer resistances.

4.5.2 Determination of the product flow rates

1. Calculate the roots Φ_F of the characteristic equation of the feed composition from equation (4.9).
2. Calculate the flow rate of the vapour in the rectifying section and the unknown product flow rates applying the method of Shiras et al. [38] as based on equation (3.43a).
3. Calculate the minimum flow ratio of the rectifying section from equation (4.10a) and the maximum flow ratio of the stripping section from equation (4.10b), respectively.

4.5.3 Determination of the distillation lines

4.5.3.1 Numerical solution of the differential equations
With the known product compositions and applying a set of the differential equation (4.14) according to the chosen distribution of the mass transfer resistances the distillation lines are calculated for any flow ratios as a function of the number of transfer units or theoretical stages. The optimal feed location follows from a total cost optimisation based on a calculation of the number of transfer units or theoretical stages vs. the reflux ratio L/D [17].

4.5.3.2 Distillation spaces and distillation lines based on the algebraic solution of the differential equations
With the known product compositions and limiting flow ratios calculated as discussed above, the eigenvalues (relative volatilities) of the distillation spaces are obtained from equation (3.43a) with the transformation matrices (4.33) defining the composition of the nodes applying equation (4.34). The composition of the distillation lines in the transformed coordinate systems is obtained from equation (4.35) vs. the number of transfer units or theoretical stages with the mole fractions of the distillation lines following from the inverted matrix (4.33).

5 Distillation of real mixtures

The vapour-liquid-equilibrium of real mixtures can be calculated by introducing a correction term into Raoult's law, the so-called activity coefficient, whereas the gas phase may still be treated like an ideal gas if the pressure is moderate with regard to the critical pressure of the components as is the case in most distillation problems. As a result of this correction the relative volatilities of the mixture are not constant anymore but become a function of the composition. Mixtures where all relative volatilities are either larger or smaller than one are called zeotropic mixtures and mixtures where one or more of the relative volatilities are equal to one are defined as azeotropic mixtures.

Zeotropic mixtures behave qualitatively like ideal mixtures but since the relative volatilities are a function of the composition the design equations must be solved numerically. In principle, however, any zeotropic system can be approximated at least qualitatively by ideal binary systems with constant relative volatilities [15, 16] with the quality of the approximation depending on the information used in calculating the constant relative volatilities of the ideal binary systems:

1. The most simple approximation of the relative volatility of a binary mixture is based on the vapour pressures of the pure components, i.e. the relative volatility of a binary mixture with the components i and k is

$$\alpha_{ik} = \frac{p_i^0}{p_k^0}\,. \tag{5.1}$$

2. A better approximation is obtained by defining a "corrected vapour pressure"

$$\overline{p_i^0(x_i = 0, \theta_i)} = \gamma_i^\infty \cdot p_i^0(\theta_i)\,, \tag{5.2}$$

$$\overline{p_i^0(x_i = 1, \theta_i)} = p_i^0(\theta_i)\,. \tag{5.3}$$

Taking the geometric mean of the ratio of the corrected vapour pressures at $x_i = 1$ and $x_i = 0$ results in

$$\alpha_{ij} = \left(\frac{\gamma_i^\infty \cdot p_i^0(\theta_j) \cdot p_i^0(\theta_i)}{p_j^0(\theta_j) \cdot \gamma_j^\infty \cdot p_j^0(\theta_i)} \right)^{0.5}\,. \tag{5.4}$$

3. Another method is to calculate the real relative volatilities of the binary mixture at different compositions of the mixture and to use an appropriate averaging procedure to determine the best approximation of the real relative volatilities by a constant relative volatility.

The effect of the mass transfer ratio on the course of real distillation lines may be accounted for by first solving the composition at the liquid–vapour-interface applying equation (3.19) and the subsequent numerical integration of equation (4.18).

https://doi.org/10.1515/9783110739732-005

Another, though rare feature of real zeotropic mixtures is the occurrence of inflection points in the course of the distillation lines due to a change in the ordering of the relative volatilities within the concentration range. The effect of such phenomenon on the feasible product domains will be discussed in the context of azeotropic mixtures.

5.1 Zeotropic mixtures

5.1.1 Binary mixtures

5.1.1.1 Enthalpy-concentration-diagram

The enthalpy-concentration-diagram is an exact representation of a real binary mixture and provides next to more realistic concentration, temperature and flow profiles in the column also information on the energy requirements of the condenser and the evaporator. This additional information is related to the difference points dp_D and dp_B given by the intersection of the q-line with the balance lines over a cross section of the rectifying section and the distillate and the intersection of the q-line with the balance lines of a cross section of the stripping section and the bottom product, respectively. The q-line is the line coinciding with the extension of the equilibrium vector $(x \rightarrow y^*)$ passing through the feed location x_F unless there is an extension of an equilibrium vector to the right or the left of the q-line yielding a difference point with a higher or lower enthalpy then the difference point related to the q-line, respectively. In such a case the limiting flow ratios are given by the difference points with the highest or lowest enthalpy.

In the enthalpy-concentration-diagram the operating lines according to equation (3.2) are represented by straight lines originating from the difference point and intersecting with the saturated vapour and the saturated liquid curve as shown in Fig. 5.1, i.e. for any flow ratio $R > R_{min}$ the difference points of the distillate and the bottom product shift to a higher and lower enthalpy, respectively, and become infinite at $R = 1$.

Like in the McCabe–Thiele-diagram the number of transfer units for any flow ratio $R_{min} \leq R \leq 1$ is calculated numerically applying equation (3.22) or (3.23) with the driving forces taken from the enthalpy-concentration-diagram with the height of the column given by the product of $NTU \cdot HTU$.

The number of theoretical stages follow again from the same procedure as discussed above in the context of the McCabe–Thiele-diagram and as shown in Fig 4.7 for the minimum reflux case. The optimal flow ratios and the optimal location of the feed are obtained again from an economic optimisation.

It should be noted that the operating line according to equation (3.2) corresponding to Fig. 4.7 is a curved line since the flow ratio is constant only if the saturated vapour line and the saturated liquid line run in parallel. The flow ratio for any concentration x follows from the mass balance line of a section, i.e. a straight line originating

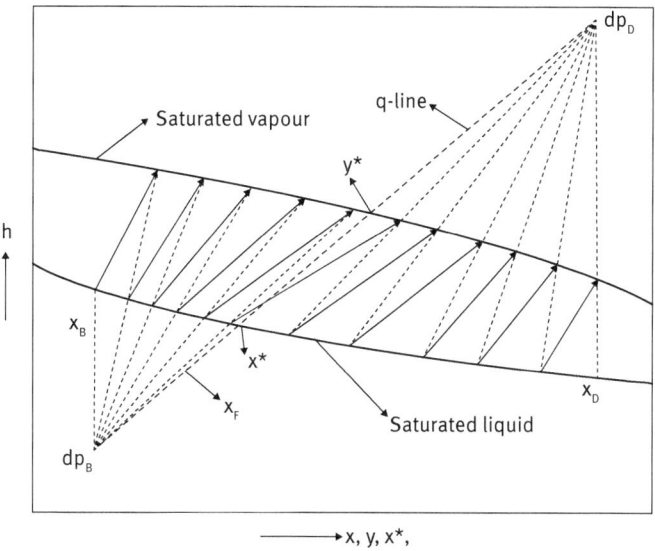

Fig. 5.1: Enthalpy-concentration-diagram (dotted lines = equilibrium vectors, dash-dotted lines = mass balance lines, dp = difference points) [39], $R = 0.63 > R_{min} = 0.5$).

from the differential point of this section and passing through x of the saturated liquid line and the saturated vapour line. The respective flow ratio is then given by the ratio of the distance from the differential point to the saturated vapour line divided by the distance from the same difference point to the saturated liquid line.

The energies removed in the condenser and added in the evaporator are given by the difference of the enthalpies of the related difference point and the product times the respective flow rate, i.e.

$$Q_c = (hd_D - h_D) \cdot D \tag{5.5}$$

and

$$Q_E = (hd_B - h_B) \cdot B. \tag{5.6}$$

In many cases the binary equilibrium can be approximated by a system with constant relative volatilities allowing to use the analytical solutions given in Chapter 4.1.

5.1.2 Ternary mixtures

The design of the distillation of zeotropic ternary mixtures follows from the same procedures as outlined in Chapter 4.2 for ideal mixtures except that the distillation lines have to be calculated numerically. It should be remembered, however, that the distillation lines are a function of the ratio of the mass transfer coefficients as shown in Chapter 3.4.2.

5.1.2.1 Distillation at total reflux

An example for a distillation at total reflux is given in Fig. 5.2 showing the binary systems of the ternary mixture acetaldehyde–methanol–water and the distillation lines at total reflux in Fig. 5.3, respectively. Except for the binary acetaldehyde–methanol which exhibits an inflection point at about $x_1 = 0.1$, the binaries behave almost like ideal systems.

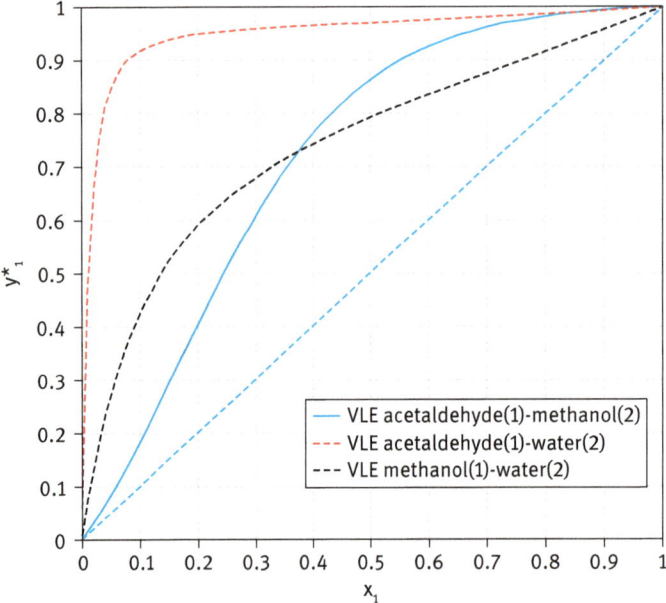

Fig. 5.2: Binary vapour–liquid-equilibria of the zeotropic mixture acetaldehyde–methanol–water at 1 bar.

Figure 5.3 also shows the distillation lines based on averaged constant relative volatilities calculated from the thermodynamic data of the mixture given in Table 5.1 [20] indicating the principle agreement between the distillation lines of ideal and zeotropic mixtures.

5.1.2.2 Distillation at partial reflux

The distillation lines of the stripping and rectifying section of the above mixture approximated as ideal mixture at limiting flow ratios are shown in Fig. 5.4.

The corresponding distillation lines of the real mixture are given in Fig. 5.5 indicating a reasonable agreement between the approximated and the real distillation lines.

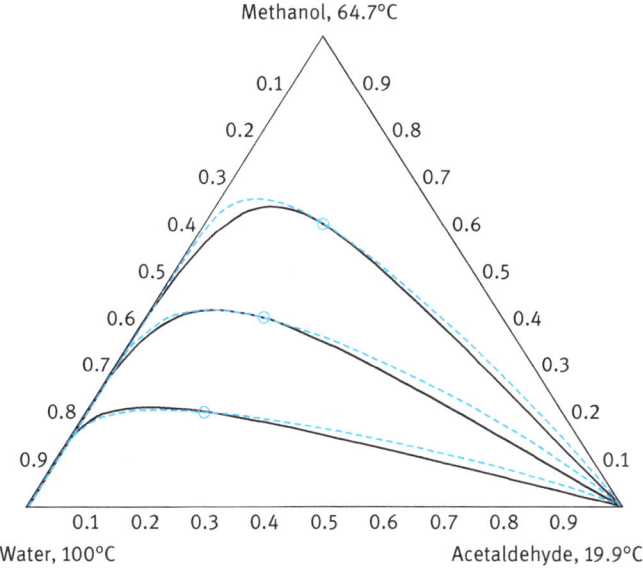

Fig. 5.3: Comparison of the real distillation lines and the distillation lines calculated with constant relative volatilities.

Tab. 5.1: Thermodynamic properties of the ternary zeotropic mixture acetaldehyde (1)–methanol (2)–water (3)

Component *i*		**1**	**2**	**3**
Boiling point at 1 bar	°C	19.9	64.7	100
Vapour pressure in bar vs. temperature	19.9	1	0.13	0.025
	64.7	4.39	1	0.24
	100	9.16	3.49	1
Limiting activity coefficients	γ_{1i}^∞	1	0.317	13.1
	γ_{2i}^∞	0.425	1	2.38
	γ_{3i}^∞	14.2	1.95	1
Limiting binary relative volatilities	α_{1i}	1	1.40	120
	α_{2i}	0.06	1	8.31
	α_{3i}	0.356	0.47	1
Av. binary volatility		$\alpha_{12} = 4.8$	$\alpha_{23} = 4.2$	$\alpha_{13} = 18.7$
Rel. ternary volatity		$\alpha_1 = 20.2$	$\alpha_2 = 4.2$	$\alpha_3 = 1.0$

5.1.2.3 Product domains and minimum flow ratio

As pointed out in Chapter 4.2.4 a determination of the product domains follows from the distillation line of the liquid through the liquid composition of the feed and the distillation line of the vapour through the composition of the vapour in equilibrium

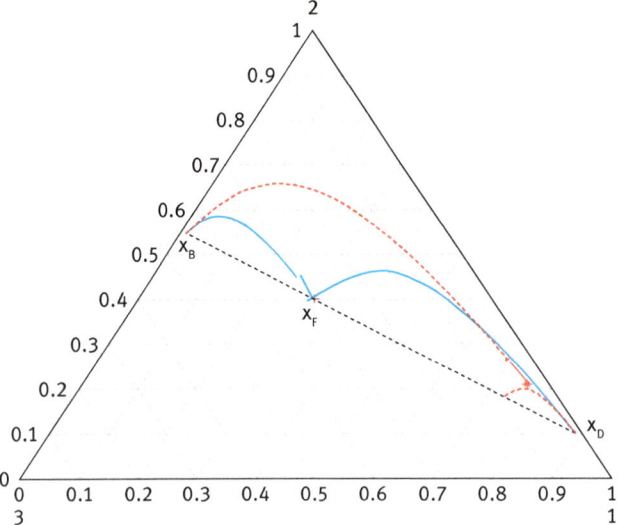

Fig. 5.4: Distillation lines of the idealized mixture acetaldehyde (1)–methanol (2)–water (3).

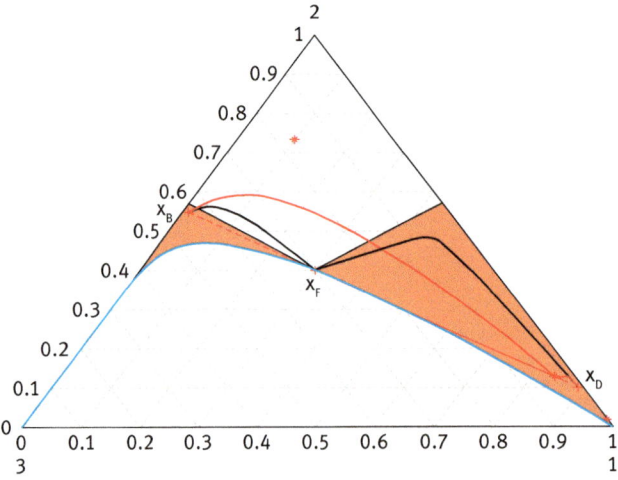

Fig. 5.5: Product domains and distillation lines of the real zeotropic mixture acetaldehyde (1)–methanol (2)–water (3).

with the composition of the liquid feed and the limiting distillation lines belonging to the feasible pure top and bottom product as shown in Fig. 5.5.

Once the product domains are known, the limiting flow ratios are obtained by trial and error starting e.g. with the assumption of sharp separations, i.e. approximately binary products and limiting flow ratios of about 1 and subsequently reducing the purity of the products as well as changing the corresponding flow ratios. Additional infor-

mation may be gained from assuming products located on the product line coinciding with the equilibrium vector of the feed as this allows to determine the corresponding limiting flow ratios applying equations (4.2a) and (4.2b).

5.1.3 Multicomponent mixtures

As in the case of ideal systems, the design procedures for the distillation of real ternary mixtures can be extended to zeotropic mixtures with any number of components by solving the resulting system of equations based on equation (4.14) or (4.37) to calculate the distillation lines or the reversible distillation lines, respectively, as well as the feasible product domains. The limiting flow ratios are obtained as discussed above.

5.2 Azeotropic mixtures

Azeotropic mixtures differ from zeotropic mixtures thereby that each azeotrope adds an additional "component" to the mixture as with respect to distillation there is no difference between a pure component and an azeotrope except that an azeotrope is next to a function of the pressure and the temperature also a function of the concentration [27, 28]. Thus, z azeotropes add z "pseudocomponents" to the mixture. Since such systems are thermodynamically unstable they disintegrate to a mixture with up to $(c - 1 + z)$ independent zeotropic subsystems [16, 40].

5.2.1 Binary mixtures

The enthalpy-concentration-diagram also allows to solve a binary distillation problem involving a binary azeotropic mixture by taking into account that the azeotropic composition is an absolute boundary which can not be crossed in distillation, i.e. it behaves like a pure component.

A solution applying the McCabe–Thiele-diagram is discussed in Fig. 5.7 showing the equilibrium curve of ethanol (1)–benzene (2) with an azeotrope (3) at $x_{1,Az} = 0.46$ [20]. Due to the azeotrope the binary mixture may be regarded as two independent binary subsystems consisting of the azeotrope (3) and the ethanol (1) and the azeotrope (3) and the benzene (2), respectively. Introducing new variables ξ and η as indicated in Fig. 5.6 for the system 1, both binary subsystems may be treated as an ideal system [15, 40].

The limiting relative volatilities of the transformed binary subsystems are identical to the slopes of the equilibrium line at the positions $x_1 = 0$; $x_1 = x_{Az}$ and $x_1 = 1$ as indicated by the dotted lines in Fig. 5.6.

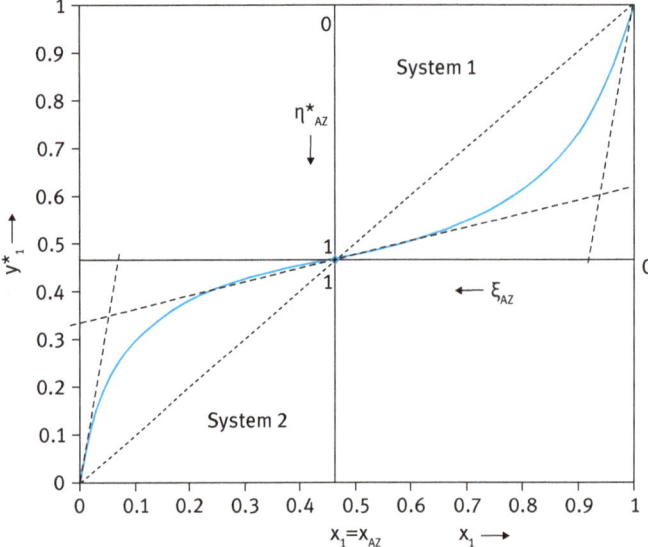

Fig. 5.6: Equilibrium curve of the binary mixture ethanol (1)–benzene (2) at 1 bar.

The graphically determined slopes are given in Table 5.2 and the approximately constant relative volatility were obtained by taking the geometric mean of the two limiting relative volatilities, i.e.

$$\alpha_{\text{Az}\,i} = \sqrt{\alpha_{\text{Az}} \cdot \alpha_i} \,. \tag{5.7}$$

Tab. 5.2: Approximated constant relative volatilities of the binary azeotropic mixture ethanol (1)–benzene (2)–azeotrope (3)

Component i	1	2	3
Limit. rel. volatility of component i	6.41	3.85	4.10
Av. bin. rel. volatility	$\alpha_{\text{Az}\,2} = 5.13$	$\alpha_{\text{Az}\,1} = 3.97$	

A comparison of the real with the approximated vapour-liquid-equilibria is given in Fig. 5.7 and the agreement is quite satisfactory.

Another approximation of the relative volatilities of the subsystems (equivalent to the method 3 above) is obtained by averaging the relative volatilities taken at several concentrations of the equilibrium line of a subsystem according to (see system 1 in Fig. 5.4)

$$\alpha_T = \frac{\eta \cdot (1 - \xi)}{\xi \cdot (1 - \eta)} \,. \tag{5.8}$$

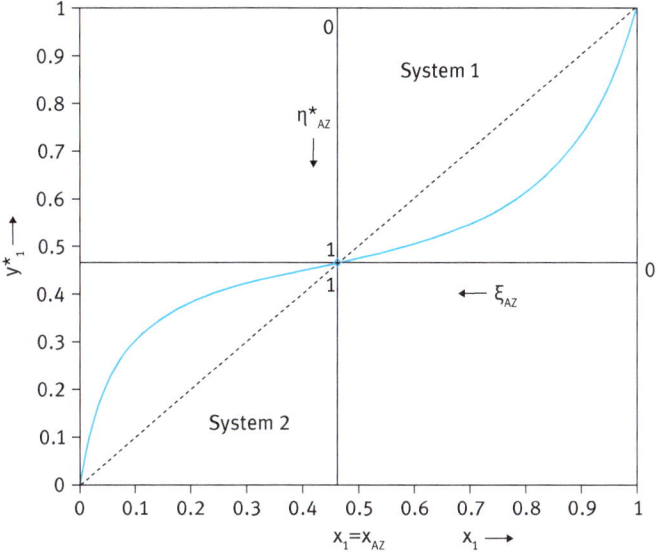

Fig. 5.7: Comparison of the real and approximated vapour–liquid-equilibria.

5.2.2 Ternary mixtures

Ternary mixtures may exhibit up to three binary and one ternary azeotrope and the principle behaviour of their distillation lines has been discussed by Konovalov [41], Reinders et al. [42], Van Dongen et. al [43], Stichlmair [44], Vogelpohl [45], Stichlmair et al. [46], Stichlmair et al. [47], Wahnschafft et al. [48], Laroche et al. [49], Fidkowsky et al. [50], Davydian et al. [51], Rooks et al. [52], Krolikowski [53], Danilov et al. [54].

Konovalov established the important rule, that in simple distillation at constant pressure the temperature of the residue increases with time and that at constant temperature the pressure increases with time. Since simple distillation and distillation at total reflux follow the same mathematical model, this rule allows to determine the principal behaviour of distillation lines for a mixture of any complexity if e.g. at constant pressure the boiling temperatures of the pure components and the azeotropes are known. The rule of Konovalov is equivalent to the statement that the distillation lines begin at the node(s) with the lowest relative volatility with respect to the relative volatility of their neighbouring node(s) and end at the node(s) with the highest volatility with respect to the relative volatility of their neighbouring node(s). It should be noted that in some cases the rule of Konovalov gives more than one solution [50]. In such a case additional information like the temperature field, the calculation of some distillation lines in the uncertain area or the theory of directed graphs [50, 52] can be applied to obtain the correct solution.

5.2.2.1 Distillation at total reflux

Since an azeotrope behaves like a pure component a ternary mixture with an azeotrope like the mixture acetone–chloroform–benzene as shown in Fig. 5.8 is strictly speaking a 4-component mixture. In order to fit into the 3-component Gibbs triangle, it has to split into two 3-component mixtures namely the acetone–azeotrope–benzene and the chloroform–azeotrope–benzene mixture separated by the additional separation line or "binary mixture" azeotrope–benzene. In contrast to the 2-component binary mixtures defining the Gibbs triangle a separation line connected with an azeotrope has 3 components and is, due to the real behaviour of an azeotropic mixture, generally curved.

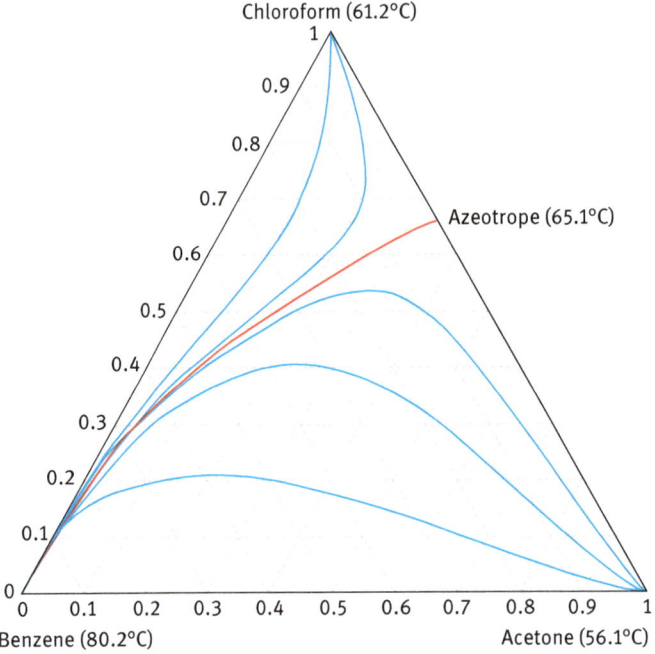

Fig. 5.8: Distillation lines of the azeotropic mixture acetone (1)–chloroform (2)–azeotrope (3)–benzene (4) at total reflux.

Since the boiling point of the binary azeotrope is higher than the boiling point of acetone and chloroform but below the boiling point of benzene, the azeotrope takes the role of an intermediate boiling component with the result that the total distillation space is separated into two independent distillation subspaces with the separation line representing a "binary mixture" common to both distillation subspaces.

An example of the distillation lines of a so-called four-sided distillation space is the mixture methyl acetate–methanol–ethyl acetate as given in Fig. 5.9.

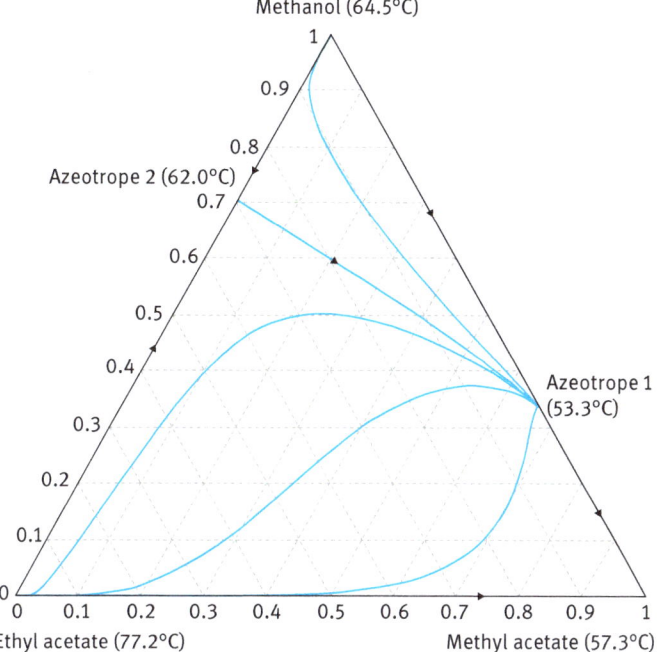

Fig. 5.9: Distillation lines of the azeotropic mixture azeotrope (1)–methyl acetate (2)–azeotrope (3)–methanol-ethyl acetate (4) at total reflux.

The structure of the total distillation field follows from the first rule of Konovalov stating that a distillation line at total reflux always proceeds in the direction of the decreasing boiling temperature as indicated by the arrows, i.e. the temperature distribution in Fig. 5.9 must result in a separation line extending from the azeotrope 2 to the azeotrope 1 creating a ternary and a so-called four-sided subspace. The distillation lines of this four-sided subspace may be regarded as a superposition of a distillation line of the ternary mixture azeotrope 1–methyl acetate–ethyl acetate and a distillation line of the ternary azeotrope 1–azeotrope 2–ethyl acetate. It should be noted that the 'virtual' separation line extending between the azeotrope 1 and the pure ethyl acetate has no effect on the field of distillation lines of the four-sided subspace.

The principle procedure of such a superposition is demonstrated in Fig. 5.10 showing the superposition of a distillation line of the upper ideal ternary mixture with a distillation line of the lower ideal ternary mixture. Starting from an assumed binary composition x_{14}, the intersections of the line $x_{14} > 2$ and the line $x_{14} > 3$ with the corresponding binaries yields the binary compositions x_{12}, x_{24}, x_{13} and x_{34}. The intersections of the lines $x_{12} > x_{34}$ and $x_{24} > x_{13}$ then yield one composition of the 'quaternary' distillation line. By changing the initial composition x_{14}, the complete 'quaternary' distillation line can be determined.

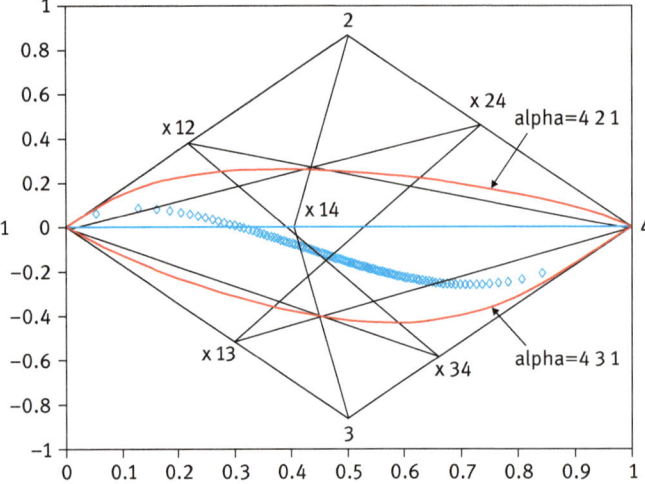

Fig. 5.10: Superpositon of two ternary mixtures to a four-sided distillation line.

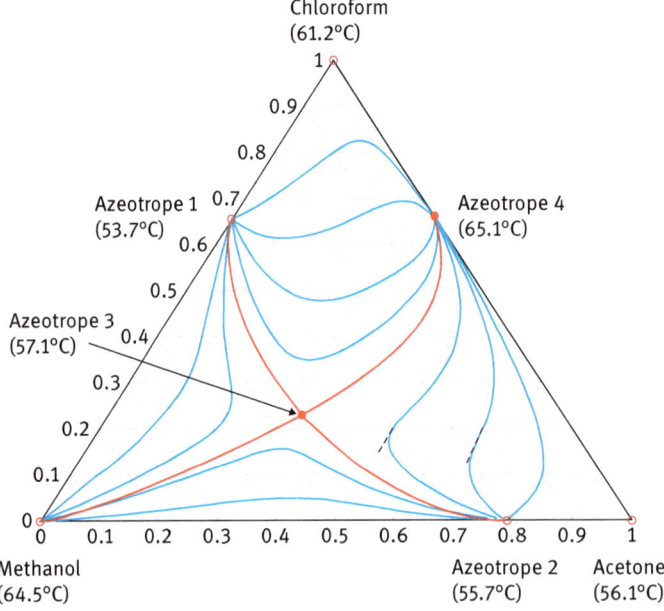

Fig. 5.11: Distillation lines of the azeotropic mixture acetone (1)–chloroform (2)–methanol (3) at total reflux.

One of the most complex azeotropic system is the mixture acetone–chloroform–methanol with three binary and one ternary azeotropes as shown in Fig. 5.11.

From the boiling temperatures of the pure components and the azeotropes follows again from Konovalov's rule that the total distillation space is split into two ternary and two four-sided subspaces. Rigorous methods to determine the structure of more complex mixtures are based e.g. on the theory of directed graphs [52].

As discussed in Chapter 3 the separation lines are the Eigencoordinates of the subspaces taking a value of 0 at the unstable node and a value of 1 at the stable node of the respective distillation subspace with each subspace representing a zeotropic subsystem. The subspaces can be approximated by ideal systems with straight separation lines as shown e.g. in Fig. 5.12 with the approximated relative volatilities calculated from the corrected vapour pressures as given in Table 5.3 [16].

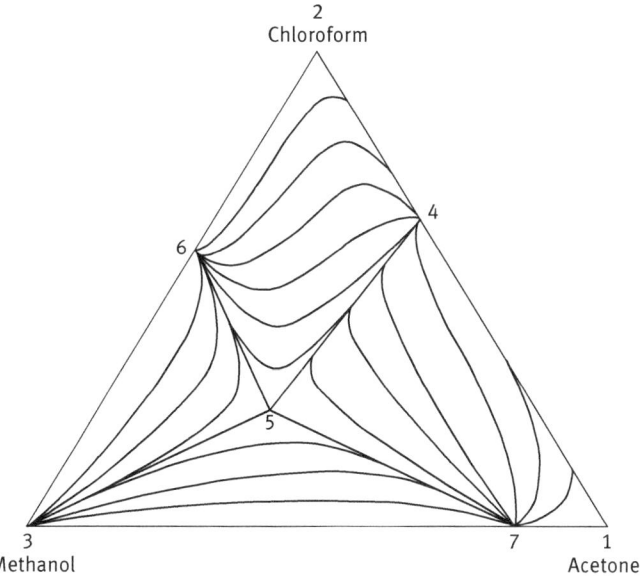

Fig. 5.12: Distillation lines of the azeotropic mixture acetone (1)–chloroform (2)–methanol (3) approximated as an ideal system [16].

Tab. 5.3: Properties of the ternary mixture acetone–chloroform–methanol

Component	Az. 1	Az. 2	Acet. 3	Az. 4	Chlor. 5	Az. 6	Metha. 7
Boiling point at 1 bar in °C	53.4	55.7	56.2	57.5	61.2	64.4	64.7
Corr. vap. press. at 60 °C in bar	1.165	1.155	1.140	1.042	0.964	0.839	0.834
Rel. volatility at 60 °C	1.400	1.385	1.367	1.249	1.156	1.006	1.000

* as calculated with equation (5.7) with reference to methanol

5.2.2.1.1 Product domains

Since the principal behaviour of the distillation lines is not affected by the distribution of the mass transfer resistances, the discussion of the product domains of real mixtures will be limited to the case of a total mass transfer resistance on the vapour side. The product domains of the reversible distillation of real mixtures are discussed in [48, 53, 55].

The product domains of a real zeotropic mixture resemble those of an ideal mixture as long as the order of the relative volatilities does not change. Any such disordering of the relative volatilities results in inflection points of the distillation lines at $L/V = 1$ producing additional product domains with one border defined by the tangent to the respective distillation line as shown in Fig. 5.13.

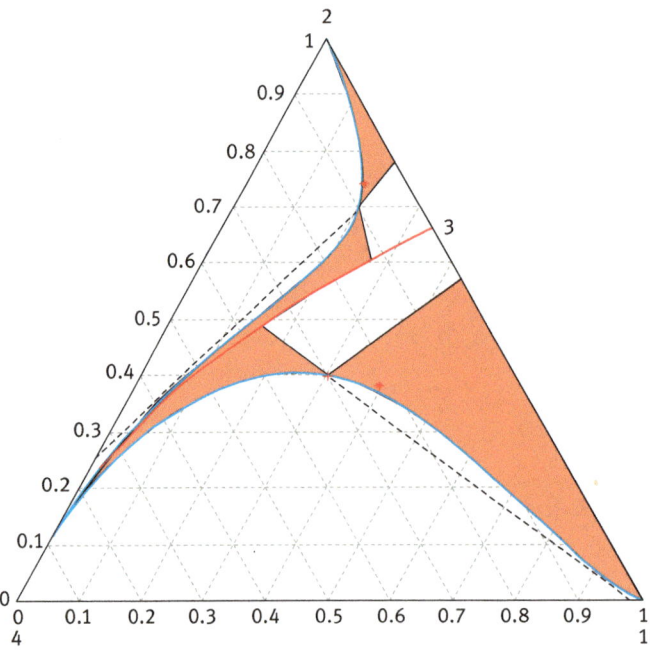

Fig. 5.13: Feasible product domains of the mixture acetone (1)–chloroform (2)–benzene (4), (filled areas = product domains, dash-dotted lines = tangents).

The additional product domains are the unfilled areas defined by the distillation lines, the tangents and the corresponding sides of the Gibbs triangle. This phenomenon of an additional product domain would allow e.g. to produce a distillate with a high concentration of acetone and a low concentration of benzene not possible in a normal distillation [50].

The largest effect of an inflection point of the distillation line on the feasible product domain is given if the feed composition coincides with the inflection point as shown e.g. in Fig. 5.14. The additional domain is defined by the distillation line of the lower subspace, the tangent to the distillation line passing through the feed composition and the side 2–5 of the Gibbs triangle.

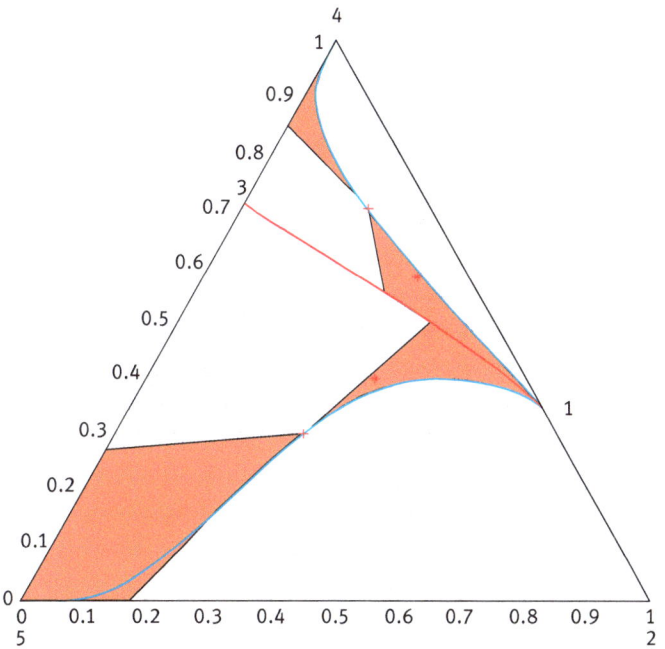

Fig. 5.14: Product domains of the lower and upper subspace of the mixture methylacetate (1)–methanol (2)–azeotrope (3)–ethylacetate (4), (x_F = [0.3 0.4 0.3], [0.2 0.7 0.1]).

It should be noted that the lower subspace 1–2–3–5 is the projection of a quaternary mixture represented by a tetrahedron (see Fig. 5.10). The corresponding separation lines (= edges of the tetrahedron) 1–5 and 2–3 are of no significance with respect to the distillation lines of the four-sided subspace.

5.2.2.1.2 Real separation lines
Whereas the separation lines defining the distillation space of a mixture with constant relative volatilities are always straight lines, the separation lines extending from an azeotrope into the composition space are generally curved due to the changing relative volatilities and thus depend on the mathematical model used to calculate the separation line as indicated e.g. in Fig. 5.15 for the mixture acetone (1)–chloroform (2)–benzene (3).

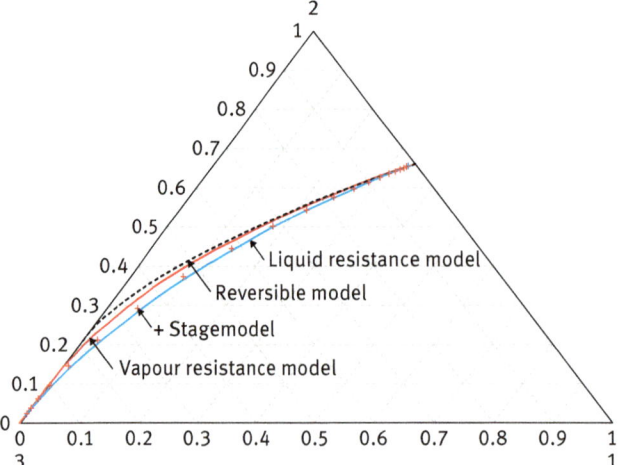

Fig. 5.15: Separation lines of the different mathematical models.

In contrast to the separation lines of the mass transfer resistance models including the theoretical stage model, the separation line designated as 'reversible model' is not a single reversible distillation line extending from the pure component 3 to the azeotrope but the envelope of the initial conditions of those reversible distillation lines ending at the azeotrope. This is due to the fact that any composition on an adiabatic distillation line may be used as an initial condition resulting in the same trajectory whereas any composition different from the initial composition of a reversible distillation line used as the initial condition will result in a different trajectory [45]. This different behaviour of the adiabatic and the reversible distillation has been discussed extensively by Wahnschafft et al. [48], Fidkowsky et al. [50] and Krolikowsky [53] who called the reversible distillation lines pinch-point curves and the reversible separation line a product pinch-point curve or a pitchfork distillation boundary, respectively.

5.2.2.2 Distillation at partial reflux

Since the distillation subspaces of an azeotropic mixture have – in principle – the properties of a zeotropic system, the distillation lines at partial reflux, the limiting flow ratios, the optimum location of the feed a.s.o. follow from the same procedures as discussed in Chapter 5.1.2 in the context of real zeotropic mixtures except for the phenomenon that the separation line extending from an azeotrope into the concentration space can be "crossed" by a distillation line under partial reflux conditions as shown by Wahnschafft [48].

5.2.2.2.1 Real distillation spaces

It should be remembered that the distillation spaces of the stripping and the rectifying section are a function of the feed and product compositions as well as the flow ratios L/V. Thus, a separation line due to an azeotropic point will also shift with the distillation spaces and possibly allow for product compositions not accessible at $L/V = 1$ as indicated e.g. in Fig. 5.16.

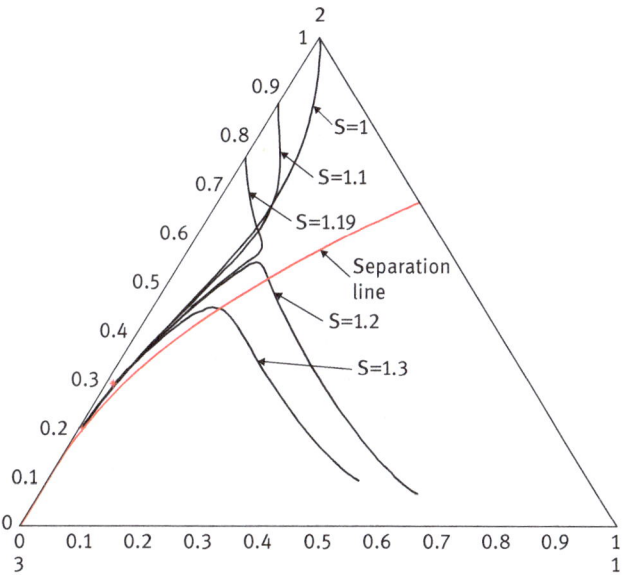

Fig. 5.16: "Crossing" of a separation line (initial composition of the distillation lines $x0 = [0.005\ 0.002\ 0.795]$).

The reason for such a "crossing" of a separation line valid at total reflux is the change of the flow ratio from $R = S = 1$ in Fig. 5.8 to flow ratios $S > 1$ and the related shift of the distillation spaces as indicated by the vector field and the distillation spaces in Fig. 5.17. The saddle point at the azeotrope in Fig. 5.8 has moved in Fig. 5.17 to $x = [0.156\ 0.538]$, the point 1 representing the pure acetone to $x = [0.607\ 0.100]$ and the point 3 representing the pure benzene to $x = [-0.005\ -0.094]$. The full and the dashed lines are the related separation lines of the liquid and the vapour distillation space, respectively. The thin full line is the separation line at $R = S = 1$.

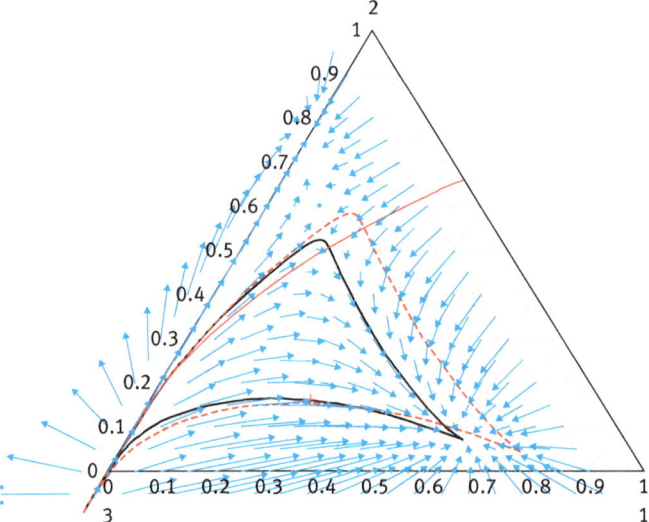

Fig. 5.17: Shift of the lower distillation space of Fig. 5.8 due to the change in the flow ratio S from 1 to 1.2.

5.2.2.2.2 Limiting flow ratios

A rigid method to determine the limiting flow ratios of real mixtures exists only for so-called sharp separations, i.e. separations where the distillate and the bottom product are free of the lowest or the highest boiling component, respectively. For a ternary mixture with the split 1–(2)/(2)–3 the corresponding limiting flow ratios are obtained either directly from the two binary mixtures or by ternary calculations. An example is given in Fig. 5.18 with the ratio of the rectifying section taken from the binary mixture acetone-chloroform as $R = 0.76$ and the flow ratio of the stripping section from the balance equation (4.10c) as $S = 1.3$.

Another approach for real systems is to approximate the real mixture by an ideal mixture as demonstrated in Chapter 5.2.3.1 and then applying the equations valid for ideal mixtures to calculate approximate limiting flow ratios or by using the graphical method based on the separation lines as discussed in Chapter 4.2.5.1.

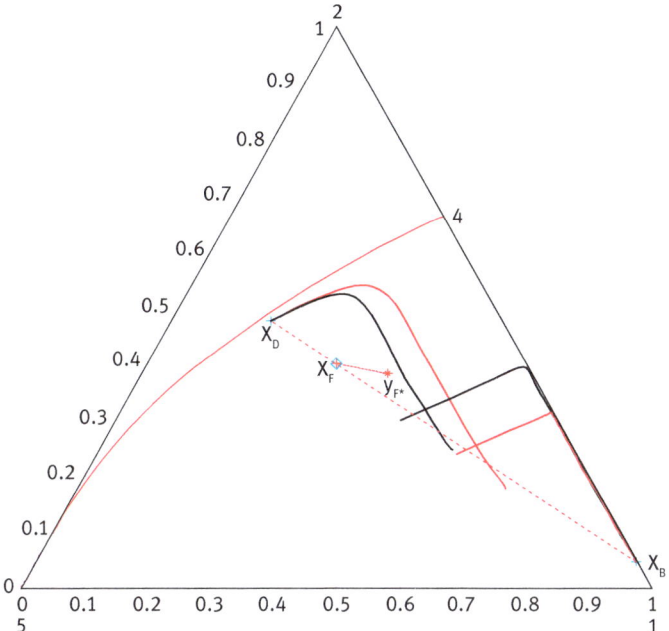

Fig. 5.18: Distillation lines of a distillation with a sharp distillate composition.

5.2.3 Multicomponent mixtures

The distillation space of multicomponent azeotropic mixtures breaks down into zeotropic subspaces which can be treated by a logical expansion of the equations discussed in the context of ternary and quaternary zeotropic mixtures.

It should be noted that the complexity of the distillation of a multicomponent mixture does not increase with the number of components as the relevant equations are a manageable number of vector equations. In contrast, the complicity of the distillation problem increases exponentially with the number of components and the solutions, therefore, are more and more difficult to visualize.

6 Computer programs

The computer subprograms used for the calculations and in designing the figures of this book are in part based on a comprehensive Graphical User Interface MATLAB® program called DISTLAB® developed by Adiche [37].

The physical data used in the subprograms are taken from the DECHEMA Series [20].

https://doi.org/10.1515/9783110739732-006

7 References

[1] Forbes, R. J.: *Of the Art of Distillation from the Beginnings up to the Death of Cellier Blumenthal*, Copyright 1948 by E. J. Brill, Leiden, Holland

[2] Underwood, A. J. V.: Distillation – Art and Science, *Chemistry and Industry*, June 23 (1963)

[3] Hausbrand, E.: *Die Wirkungsweise der Rectificir- und Destillirapparate*, Berlin (1893) cited in [1]

[4] Sorel, E.: *La rectification de l'alcool*, Paris (1894) cited in [1]

[5] Lewis, W. K., *Trans. Am. Inst. Chem. Engrs.*, 44 (1922) 329

[6] Lord Rayleigh, *Philos. Magazine*, 4 (1902) 521

[7] Fenske, *Ind. Eng. Chem* 24 (1932) 482

[8] Underwood, A. J. V., *Trans. Inst. Chem. Engrs.*, 10 (1932) 112

[9] Underwood, A. J. V.: *Fractional distillation of ternary mixtures*, Part I, (1945) 111–118, Part II (1946) 598–613

[10] Underwood, A. J. V.: Fractional distillation of multi-component mixtures – Calculation of minimum reflux ratio, *J. Inst. Petroleum*, 32 (1946) 614–626

[11] Hausen, H.: Rektifikation von Dreistoffgemischen, *Forschung Gebiet Ing.-Wesen* 6 (1935) 9–22

[12] Hausen, H.: Rektifikation idealer Dreistoff-Gemische, *Z. angew. Physik*, 2 (1952) 41–51

[13] Vogelpohl, A. Rektifikation idealer Vielstoffgemische, *Chem.-Ing.-Tech.* 42 (1970) 1377–1382

[14] Vogelpohl A.: A unified description of the distillation of ideal mixtures, *Chemical Engineering & Processing*, 38 (1999) 631–634.

[15] Vogelpohl, A.: Die naeherungsweise Berechnung der Rektifikation von Gemischen mit azeotropen Punkten, *Chem.-Ing.-Tech.* 46 (1974) 195

[16] Vogelpohl A.: On the Relation between Ideal and Real Mixtures in Distillation, *Chem. Eng. Technol.* 25 (2002) 869–872

[17] King, C.J,: *Separation Processes*, 2nd Edition, McGraw-Hill Book Company (1980)

[18] Westerberg, A. W.: The Synthesis of Distillation-Based Separation Systems, *Computers and Chemical Engineering*, 9 (1985) 421–429

[19] Hausen, H.: Berechnung der Rektifikation mit Hilfe kalorischer Mengeneinheiten, *Z. VDI-Beiheft Verfahrenstechnik* (1942) 17–20

[20] Gmehling, J., V. Onken, W. Arlt: Vapour–liquid Equilibrium Data Collection, *Dechema Chemistry Data Series*, Frankfurt/M., since 1977

[21] Krishna, R., G. L. Standart: A Multicomponent Film Model Incorporating a General Matrix Method of Solution to the Maxwell-Stefan-Equations, *AICHE J.* 22 (1976) 383–389

[22] Whitman, W. G., *Chem. Met. Engr.* 29 (1923) 147

[23] Chilton, T. H., A. P. Colburn: Distillation and Absorption in Packed Columns, *Ind. Eng. Chem.* 27 (1935) 255

[24] Rische E. A.: Rektifikation idealer Gemische unter der Voraussetzung, dass der Widerstand des Stoffaustausches allein auf der Flüssigkeitsseite liegt, *Z. angew. Phys.* 7 (1955) 90–96

[25] Korn, G. A., T. M. Korn: *Mathematical Handbook for Scientists and Engineers*, McGraw-Hill Book Company (1968) 247

[26] Ostwald, W.: Dampfdrucke ternärer Gemische, *Abh. Math.-phys. Classe Sächs. Ges. Wiss.* 25 (1900) 413–453

[27] Vogelpohl, A.: Offene Verdampfung idealer Mehrstoffgemische, *Chem.-Ing.-Tech.* 37 (1965) 1144–1146

[28] Hausen, H. Verlustfreie Zerlegung von Gasgemischen durch umkehrbare Rektifikation, *Zeitschr. f. techn. Physik* (1932) 271–277

[29] Hausen, H.: Zur Definition des Austauschgrades von Rektifizierböden bei Zwei- und Dreistoff-Gemischen, *Chem.-Ing.-Tech.* 10 (1953) 395–397

[30] McCabe, W. L., E. W. Thiele, *Ind. Eng. Chem.* 17 (1925) 605

https://doi.org/10.1515/9783110739732-007

[31] *Bubble Tray Design Manual*, American Institute of Chemical Engineers (1958)

[32] Franklin, N. L.: The Interpretation of Minimum Reflux Conditions in Multi-Component Distillation, *Trans. I. Chem. E.* 31 (1953) 363–388

[33] Franklin, N. L.: The Theory of Multicomponent Countercurrent Cascades, *Chem. Eng. Res. Dev.* 66 (1988) 65–74

[34] Petlyuk, F. B.: *Distillation Theory and Its Application to Optimal Design of Separation Units*, Cambridge Series in Chemical Engineering, Cambridge University Press (2004),

[35] Vogelpohl, A.: Der Einfluss der Stoffaustauschwiderstände auf die Rektifikation von Dreistoffgemischen, *Forsch. Ing.-Wes.* 29 (1963) 154–158

[36] Franklin, N. L., M. B. Wilkinson: Reversibility in the Separation of Multicomponent Mixtures, *Trans. I. Chem. E.* 60 (1982) 276–282

[37] Adiche, C.: Contribution to the Design of Multicomponent Homogeneous Azeotropic Distillation Columns, *Fortschritt-Berichte VDI* 3, Nr. 881 (2007)

[38] Shiras, R. N., D. N. Hanson, C. H. Gibson: Calculation of Minimum Reflux in Distillation Columns, *Ind. Eng. Chem.* 42 (1950) 871–876

[39] Ponchon, M.: Etude graphique de la distillation fractionée industrielle, *Technique Moderne* 13 (1921) 20, 55

[40] Andersen, N. J., M. F. Doherty: An Approximate Model for Binary Azeotropic Distillation Design, *Chem. Eng. Sci.* 1 (1994) 11–19

[41] Konovalov, D.: Über die Dampfspannungen der Flüssigkeitsgemische, *Wied. Ann. Physik* 14 (1881) 34–52

[42] Reinders, W., C. H. de Minjer; Recueil Trav Chim., 59 (1940) 207, 369, 392

[43] Van Dongen, M. F. Doherty: Design and Synthesis of Homogeneous Azeotropic Distillations, *Ind. Eng. Chem. Fundam.* 24 (1985) 454–463

[44] Stichlmaier, J.: Zerlegung von Dreistoffgemischen durch Rektifikation, *Chem.-Ing.-Tech.* 10 (1988) 747–754

[45] Vogelpohl, A.: Rektifikation von Dreistoffgemischen, Teil 1: Rektifikation als Stoffaustauschvorgang und Rektifikationslinien idealer Gemische, *Chem.-Ing.-Tech.* 36 (1964) 9, 907–915, Teil 2: Rektifikationslinien realer Gemische und Berechnung der Dreistoffrektifikation, *Chem.-Ing.-Tech.* 36 (1964) 10, 1033–1045

[46] Stichlmair J, R. Fair, J. L. Bravo: Separation of azeotropic mixtures via enhanced distillation, *Chem. Eng. Prog.* (1989) 63–69

[47] Stichlmair, J. G., J. R. Herguijuela: Distillation Processes for the Separation of Ternary Zeotropic and Azeotropic Mixtures, *AICHE J.* 38 (1992) 1523–1535

[48] Wahnschafft, O. W., J. W. Koehler, E. Blass, A. W. Westerberg: The product composition regions of single feed azeotropic distillation columns, *Ind. Eng. Chem. Res.* 31 (1992) 2345–2362

[49] Laroche, L., N. Bekiaris, HW. Andersen, M. Morari: Homogeneous azeotropic distillation: Separability and synthesis, *Ind. Eng. Chem. Res.* 31 (1992) 2190–2209

[50] Fidkowski, Z. T., M. F. Doherty, M. F. Malone: Feasibility of separations for distillation of nonideal ternary mixtures, *AICHE J.* 39 (1993) 1303–1321

[51] Davydian, A. G., M. F. Malone, M. F. Doherty, Theor. Found. Chem. Engng. 31 (1997) 327–338

[52] Rooks, R. E., Vivek J., M. F. Doherty., M. F. Malone M: Structure of Distillation Regions for Mullticomponent Azeotropic Mixtures, *AICHE J.* 44 (1998) 6, 1382–1388

[53] Krolikowski, L. J.: Determination of Distillation Regions for Non-Ideal Ternary Mixtures, *AICHE J.* 52 (2006) 532–544

[54] Danilov, R. Yu., F. B. Petlyuk, L. A. Serafimov: Minimum-Reflux Regime of Simple Distillation Columns, *Theor. Found. Chem. Engng.* 41 (2007) 371–383

[55] Vogelpohl, A.: Die Produktbereiche der Vielstoffrektifikation, *Chem.-Ing.-Tech.* 6 (2012) 868–874

[56] Heaven, D. L., cited in [17]

[57] De Haan, A., H. Bosch: *Industrial Separation Processes*, De Gruyter (2013)

8 Nomenclature

Symbol	Definition	Dimensions[†]
a	Interfacial area per unit volume	L^{-1}
a_i	constant defined by equation (3.2)	–
A	Mass transfer area	L^2
A_S	Cross sectional area	L^2
B	Bottom product flow rate	M/T
c	Integration constant, number of components	–
d	Differential operator	–
d	Flow rate of a component in the distillate	M/T
D	Distillate flow rate	M/T
e	Exponent defined by equation (3.5)	–
E_i	i th Eigenvalue defined by equation (3.42)	–
E	Mole averaged relative volatility defined by equation (3.10)	–
f	Flow rate of a component in the feed	M/T
$f(x)$	Function of x	–
F	Feed flow rate	M/T
F	Cross section	L^2
h	Specific free energy	J/M
h	Length coordinate	L
hd	Difference point	–
H	Height of a packing	L
H^*	Dimensionless height of a packing	–
HTU	Height of a Transfer Unit	L
k	Ratio of the mass transfer coefficients	–
k_V, k_L	Mass transfer coefficient	$M/(L^2 \cdot T)$
ln	Natural logarithm	–
L	Liquid flow rate	M/T
L_S	Liquid holdup in Simple Distillation	M
nc	number of components	–
n_i	Flux of component i across interface	$M/(T \cdot L^2)$
NTS	Number of Theoretical Stages	–
NTU	Number of Transfer Units	–
p	Total pressure	F/L^2
P	Potential function	–
Q	State of aggregation, 1 = liquid, 0 = vapour	–
q	Thermal condition of the feed defined by equation (3.3)	–
r	Reflux ratio = L/D	–
rec	Recovery of a component in the product	
r_{in}	Mole fraction of component i in the liquid at the node n	–
R	Flow ratio rectifying section = L/V	–
s	Reboil ratio = V/B	–
s_{in}	Mole fraction of component i in the vapour at the node n	–
S	Flow ratio stripping section = L/V	–
Ss	Number of distillation sequences	–
T	Kelvin temperature	D

https://doi.org/10.1515/9783110739732-008

u	Slope of the separation lines	–
V	Vapour flow rate	M/T
x	n-dimensional vector indicating the composition of a liquid	–
x_i	Mole fraction of component i in the liquid	–
X_i	Mole ratio defined by equation (4.8)	–
y	n-dimensional vector indicating the composition of a vapour	–
y_i	Mole fraction of component i in the vapour	–
$||$	Vector	

Greek

α	Relative volatility of a component with respect to the highest boiling component	–
λ	Integration constant	–
ε	Generalized relative volatility of a component defined in Table 5.1	–
Θ	Generalized mole ratio defined in Table 5.1	–
γ	Activity coefficient of a component	–
ρ	Density	M/L³
τ	Temperature in degree Celsius	D
η	Transformed mole fraction of a component in the vapour	–
π	Corrected vapour pressure defined by equation (5.2)	F/L²
κ	Ratio of the mass transfer coefficients	–
ξ	Transformed mole fraction of a component in the liquid	–
Φ	Root of equation (3.9)	
σ	Mass transfer resistance ratio	–
Σ	Sum over all components	–
Ξ	Mole ratio defined by equation (4.50)	–
ν	Defined by equation (5.7)	–
ω	Defined by equation (5.6)	–
∞	At infinite dilution	–

Subscript

av	Average
Az	Azeotrope
B	Bottom
D	Distillate
E	Equilibrium
F	Feed
V	Vapour
i	Component
ij	related to the components i and j
in	Component i at the node n
I	Interface
j	Component, intermediate boiling component
k	Highest boiling component, number of components
key	Key component
L	Liquid
M	Mixing
min	Minimum value
P	Product

n	Node
r	Reference component
R	Rectifying section
S	Stripping section
TS	Theoretical stage
0	Initial concentration
1	Lowest boiling component

Superscript

0	Pure component
*	Equilibrium with prevailing value in the other phase
∞	Limiting value

Dimension

D	Degrees, temperature
F	Force
L	Length
M	Mole
T	Time
J	Energy

9 Glossary

Characteristic function	Equation (3.43)
Distillation line	Concentration profile of the liquid or the vapour phase in a mass transfer section
Distillation space	Polyhedron defined by the nodes and the separation lines or separation surfaces, contains the distillation lines
Distillation subspace	Subspace of a distillation space with one stable and one unstable node
Equilibrium line	Straight line coinciding with an equilibrium vector
Equilibrium vector	Straight line passing from a liquid composition to the vapour composition in equilibrium with the liquid composition
Eigencoordinates	Define the coordinates of a distillation space
Eigenvalue	Mole averaged relative volatility, root of the characteristic function, defines the composition of the nodes
Feed line	Straight line representing the mixing of the feed with the flows in the feed section
Feed section	Section of a distillation column without mass transfer where the feed stream is added to the column
Flow ratio	L/V
Key components	Components defining the split
Node	Intersection of the equilibrium and the mass balance equations for all components, composition where the driving forces of all components are zero, represents one corner point of the distillation space
Mass transfer section	Section of a distillation column with mass transfer
Partial reflux	Distillation at different flow rates of the liquid and the vapour phase
Product domain	Space defining feasible product compositions
Product line	Mass balance line connecting the products and passing through the feed composition
Product section	Section of a distillation column without mass transfer where a product stream is withdrawn from the column
Reboil ratio	V/B
Reflux ratio	L/D
Residue curves	Concentration profile of the liquid in simple distillation or the distillation at Total reflux
Separation line, separatrix	Distillation line connecting two nodes, represents a boundary of the distillation space
Sharp separation	Separation where at least one component of the mixture does not appear in the distillate or the bottom product, respectively
Simple distillation	Discontinues distillation from a still pot
Split	Defines the separation of a mixture into the distillate and bottom fraction
Subregion	see: Distillation subspace
Total reflux distillation	Distillation at equal flow rates of liquid and vapour
Theoretical stage	Mass transfer section characterised by a thermodynamic equilibrium of the outgoing flows

https://doi.org/10.1515/9783110739732-009

A Appendices

A.1 Coordinate transformation

The basic differential equation for the concentration profiles in distillation arrives from equation (3.20) written for the components i and j and with equation (3.2) and division follows

$$\frac{dx_i}{dx_j} = \frac{y_i^* - y_i}{y_j^* - y_j} = \frac{y_i^* - (a_i + R \cdot x_i)}{y_j^* - (a_j + R \cdot x_j)} \ . \tag{A.1}$$

Insertion of the equilibrium equation (3.8) yields

$$\frac{dx_i}{dx_j} = \frac{\alpha_i \cdot x_i - E \cdot (a_i + R \cdot x_i)}{\alpha_j \cdot x_j - E \cdot (a_j + R \cdot x_j)} = \frac{\alpha_i \cdot x_i - E \cdot a_i - E \cdot R \cdot x_i}{\alpha_j \cdot x_j - E \cdot a_j - E \cdot R \cdot x_j} \ . \tag{A.2}$$

The denominator in equation (3.8)

$$E = \sum_i \alpha_i \cdot x_i \tag{3.10}$$

transforms on the basis of the matrix equation (3.38)

$$|x| = |r| \cdot |\xi| \tag{3.38}$$

to

$$E = \sum_n \left(\sum_i \alpha_n \cdot r_{in} \cdot \xi_i \right) \ . \tag{A.3}$$

Replacing (see equation (3.40))

$$r_{in} = (a_i + R \cdot r_{in}) \cdot E_n \tag{A.4}$$

and taking into account

$$\sum_i (a_i + R \cdot r_{in}) = 1 \tag{A.5}$$

yields

$$E = \sum_i E_i \cdot \xi_i \ . \tag{A.6}$$

Replacing the $|x|$ on the right side of equation (A.2) by the matrix equation (3.36) and inserting equation (A.6) gives

$$\frac{dx_i}{dx_j} = \frac{\sum_n (\alpha_i \cdot r_{in} - a_i \cdot E_i - R \cdot E \cdot r_{in}) \cdot \xi_i}{\sum_n (\alpha_j \cdot r_{jn} - a_j \cdot E_j - R \cdot E \cdot r_{jn}) \cdot \xi_j} \ . \tag{A.7}$$

With equation (3.10) follows

$$\frac{dx_i}{dx_j} = \frac{r_{in} + \sum_k \left(r_{ik} \cdot \frac{E_k - E}{E_n - E} \cdot \frac{\xi_k}{\xi_n} \right)}{r_{jn} + \sum_k \left(r_{jk} \cdot \frac{E_k - E}{E_n - E} \cdot \frac{\xi_k}{\xi_n} \right)} \ . \tag{A.8}$$

Differentiation of the equation (3.36) and insertion in equation (A.8) yields

$$\frac{dx_i}{dx_j} = \frac{r_{in} + \sum_k \left(r_{ik} \cdot \frac{d\xi_k}{d\xi_n} \right)}{r_{jn} + \sum_k \left(r_{jk} \cdot \frac{d\xi_k}{d\xi_n} \right)} \ . \tag{A.9}$$

A comparison of equations (A.7) and (A.8) finally results in

$$\frac{d\xi_i}{d\xi_j} = \frac{(E_i - E) \cdot \xi_i}{(E_j - E) \cdot \xi_j} \tag{A.10}$$

https://doi.org/10.1515/9783110739732-010

or by defining an equilibrium equation in analogy to equation (3.8)

$$\eta_i^* = \frac{E_i \cdot \xi_i}{E} \tag{A.11}$$

with

$$E = \sum_i E_i \cdot \xi_i \tag{A.12}$$

results in

$$\frac{d\xi_i}{d\xi_j} = \frac{\eta_i^* - \xi_i}{\eta_j^* - \xi_j} \tag{A.13}$$

with the solution in analogy to equation (4.16)

$$\frac{\xi_j}{\xi_i} = \left(\frac{\xi_k}{\xi_i}\right)^{\frac{E_i - E_j}{E_i - E_k}} \tag{A.14}$$

with the constant λ_j determined by the initial conditions of the transformed concentrations $|\xi|$.

For a polynary mixture of n components there are $(n-1)$ independent equations of the form (A.14). The concentrations $|\xi|$ again (see Chapter 4.2) are calculated using the mass balance equation

$$\sum_i \xi_i = 1 \tag{A.15}$$

in the form

$$\frac{1}{\xi_i} = \sum_k \frac{\xi_k}{\xi_i} . \tag{A.16}$$

The concentration profiles $\xi_i = f(H)$ follow from

$$\left(\ln c_{ji} - \ln \xi_i + \frac{E_i}{E_j - E_i}(\ln \xi_j - \ln \xi_i)\right) \cdot HTU = H \tag{A.17}$$

in analogy to equation (4.35).

Application of the matrix equation (3.38) finally yields the concentrations $|x|$.

In the same way as discussed in Chapter 3 the solutions of equation (A.14) may be visualized as distillation lines within the distillation space in form of a polyhedron formed by the nodes and the separation lines or separation planes for a ternary or a polynary mixture, respectively.

A.2 Distillation of an ideal five-component mixture (King [17])

A.2.1 Conceptual design

$$F = 1;$$

Feed: $x_F = [0.05\ 0.10\ 0.30\ 0.50\ 0.05];$

$$\alpha = [3\ 2.1\ 2\ 1\ 0.8];$$

$$q = 1;$$

Φ_F-function: $\quad 1 - q = \sum_i \frac{\alpha_i \cdot x_{F,i}}{\alpha_i - \Phi_F} . \tag{A.18}$

$$(1 - q) = 0 = \frac{3 \cdot 0.05}{3 - \Phi} + \frac{2.1 \cdot 0.10}{2.1 - \Phi} + \frac{2 \cdot 0.30}{2 - \Phi} + \frac{1 \cdot 50}{1 - \Phi} + \frac{0.8 \cdot 0.05}{0.8 - \Phi};$$

$$\Phi = [2.8788\ 2.0749\ 1.3859\ 0.8118];$$

$$\text{Split} = 1\text{--}2\text{--}(3)/(4)\text{--}5;$$

$$\Phi_{\text{keys}} = 1.3859;$$

Recovery of the key components: $\text{rec}_i = x_{D,i}/x_{F,i} \cdot D/F = d_i/f_i;$

$$\text{rec}(3) = 0.9, \quad \text{rec}(4) = 0.1;$$

$$V_{\min} = \sum_i \frac{\alpha_i \cdot D \cdot x_{D,i}}{\alpha_i - \Phi_F} = \sum_i \frac{\alpha_i \cdot d_i}{\alpha_i - \Phi_F}; \tag{4.10a}$$

The unknown variables d_1, d_2, d_5 and $V_{\min}$ are determined by solving the four equations (4.10a) written with the four solutions Φ_F of equation (4.9). The set of the four equations yields a recovery of the component 1 exceeding the amount of the component 1 in the feed indicating that this component is non-distributing. Thus the flow rate of the component 1 in the distillate is set $d(1) = 0.05$, the first equation is dropped and the procedure repeated. Similarly, also the component 5 proves non-distributing leaving a solution of the last two equations as $V_{\min} = 1.132$, $d_2 = 0.099$, a distillate flow rate of $D/F = 0.47$ and a minimum flow ratio of $(L/V)_{\min} = 0.584$ as calculated by the method of Shiras et al. [38]. The concentration profiles of the liquid and the vapour are given in Fig. A1 as a function of the Number of Transfer Units.

The key components 3 and 4 intersect at the 15th NTU in good agreement with the feed conditions and the feed should be introduced at this position, therefore. Since the method of Shiras indicated that the lowest and the highest boiling component are not distributing, the calculation is extremely sensitive to the chosen concentration of the lowest boiling component in the bottom product and the highest boiling component in the distillate, respectively. The product concentrations applied in Fig. A.1 are:

$$x_B = [0.0000004\ 0.0096\ 0.0569\ 0.8528\ 0.0807]$$

$$X_D = [0.1066\ 0.2090\ 0.5757\ 0.1065\ 0.0022].$$

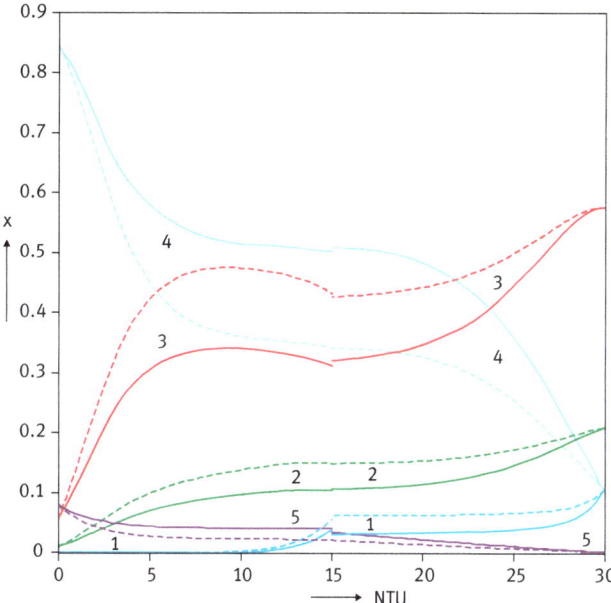

Fig. A.1: Concentration profiles of the liquid (—) and the vapour (- -) at limiting flow conditions vs. the number of transfer units ($R = R_{\min} = 0.5842$)

The dimensions of the column follow from a cost optimisation based on a calculation of the Number of Transfer Units vs. the reflux ratio. Since the investment costs decrease with an increasing reflux ratio and the energy costs increase with an increasing reflux ratio, the total costs must show a minimum vs. the reflux ratio [57].

A.2.2 Geometry of the distillation column [17, 57]

Assuming an average height of a transfer unit (HTU) = 0.4 m results in a height of the column $H = NTU \cdot HTU$ = 12.00 m with the feed introduced at a height of 6 m.

The diameter of the column follows from fluid dynamic and economic considerations based e.g. on the

- physical and chemical properties of the mixture,
- kind of internals like stages or packings,
- maximum permissible vapour and liquid load of the column,
- ease of operation in case of flow fluctuations,
- economic optimisation.

The final design of the column should be based on specific information given by the supplier of the internals.

Index

Alfons Vogelpohl
Distillation
De Gruyter Textbook